NEW DIMENSIONS IN AGRICULTURAL GEOGRAPHY

VOLUME SIX

HUMAN PERCEPTION AND TECHNOLOGICAL CHANGE IN AGRICULTURE

Concept's International Series in Geography

Volume 1 **Historical Dimensions of Agriculture**
Volume 2 **The Ecology of Agricultural System**
Volume 3 **Socio-economic Dimensions of Agriculture**
Volume 4 **Landuse and Agricultural Planning**
Volume 5 **Spatial Dimensions of Agriculture**
Volume 6 **Human Perception and Technological Change in Agriculture**
Volume 7 **Dynamics of Agricultural Development**
Volume 8 **Anthropogenic Dimensions in Agriculture**

Concept's International Series in Geography No. 4

NEW DIMENSIONS IN AGRICULTURAL GEOGRAPHY

VOLUME SIX

HUMAN PERCEPTION AND TECHNOLOGICAL CHANGE IN AGRICULTURE

NOOR MOHAMMAD
Editor

CONCEPT PUBLISHING COMPANY, NEW DELHI 110059

ISBN 81-7022-400-4 (Set)
ISBN 81-7022-406-3 (Vol. 6)

First Published 1992

Published by
Ashok Kumar Mittal
Concept Publishing Company
A/15-16, Commercial Block, Mohan Garden
NEW DELHI 110059 (India)

Lasertypeset by
Microtech Advance Printing Systems (Pvt.) Ltd.
H-13, Bali Nagar,
NEW DELHI 110015

Printed at
Sunil Printers
CB 386, Ring Road, Naraina
NEW DELHI 110028

PREFACE

AGRICULTURE CONTINUES to be the key industry. It is not only the dominant form of employment but also forms the basis for a way of life for, perhaps, two-thirds of the world's households. Agriculture being the function of physical, socio-institutional and techno-economic factors which are dynamic in nature, keeps on changing with the basic objective of increasing production and generation of foodgrains surplus. The traditional system of farming is in the process of transformation to modern agriculture which itself is undergoing a set of equally and generally even in more rapid changes. So it has added a new dimension in agriculture which needs to be studied and analysed.

A project was, therefore, taken up to accomplish this task. Original research papers dealing with various facets of agriculture were invited from scholars of different discipline from all over the world. The response was very encouraging and a large number of original papers were received. Besides these, some papers which have classical and everlasting value have also been picked up from my earlier publication – *Perspectives in Agricultural Geography*. The papers have been classified according to their basic themes and have been put into following eight volumes so as to give a systematic picture of various themes. The articles dealing with concepts, theories and models have been kept first followed by those dealing with the same theme at macro, meso and micro levels respectively.

The First Volume – *Historical Dimensions of Agriculture* – containing 19 articles deals with the origin, diffusion and development of agriculture, research trends in agricultural geography, world agricultural systems and methods and techniques in agricultural geography.

The Second Volume – *The Ecology of Agricultural System* – deals with the selected parameters of physical environment specially the climatic elements which affect the agricultural landscape in spite of technological advancement in agriculture.

The Third Volume – *Socio-economic Dimensions of Agriculture* –

containing 19 articles deals with socio-economic environment which have bearings on agricultural systems. The first 8 articles deal with social parameters specially labour, caste, women etc., while the last 11 articles examine the impact of techno-economic factors specially the various facets of irrigation which is a basic agricultural input.

The Fourth Volume – *Landuse and Agricultural Planning* – contains 19 articles dealing with concepts, theories, models of landuse, land capability and suitability, wastelands and dynamics of landuse.

The Fifth Volume – *Spatial Dimensions of Agriculture* – includes articles on agricultural systems of specific areas such as Himalayas, Bastar, Uttara Khand etc. It also deals with dynamics of crop landuse, cropping pattern and its diversification, crop ecology and crop-combination analysis.

The Sixth Volume – *Human Perception and Technological Change in Agriculture* – contains only 15 articles and covers four basic themes. The first set of 3 articles deals with human perception and his consequential response, while the second set includes articles on correlates of technological change and adoption of agricultural innovations. The third set of articles deals with the assessment of socio-economic transformation in rural India. The last set deals with agriculture and human settlements.

The Seventh Volume – *Dynamics of Agricultural Development* – includes articles on concept of development and its various facets. Besides assessing the trends of levels of agricultural development, it also examines and evaluates its impact on environment and ecological parameters.

The Eighth Volume – *Anthropogenic Dimensions of Agriculture* – containing 17 articles deals with regional imbalances in agricultural productivity, methods of its measurement, and also its correlates and determinant. It also focuses on food availability and food security at regional and continental level.

The collection of papers in eight volumes of *New Dimensions in Agricultural Geography* is designed to encapsule the major themes and recent trends and concepts in the field of agricultural geography with particular reference to conceptual revolution in the discipline. All the volumes contain a brief introduction which gives a glimpse into the topics dealt with in each of them.

The variety of articles contributed by the renowned scholars of different disciplines has made the character of these volumes multi-disciplinary and has enhanced their value. These volumes have been designed to help not only the students of a particular discipline but also to general students who wish to keep abreast of the latest developments in the field of agricultural geography. From these points of view these volumes hold a promise of rich fare and in addition to being heuristic, will provide a mine of information for discerning scholars. It is hoped that this unique collection of research papers will not only be valuable to the researchers in the fields of agricultural geography, economics, sociology, history, and agricultural science, but will also provide guidelines to decision-makers and planners.

The completion of this project has been possible due to the constant inspiration and unstinting co-operation of a host of people too numerous to acknowledge individually. However, I owe a debt of gratitude to my mentor — Dr. Mohammad Shafi, Professor Emeritus, Department of Geography, A.M.U., Aligarh who gave me valuable comments and suggestions ever since the inception of this project. Prof. R. Ramachandran and my other colleagues who offered sincere comments on the title of volumes and the collection of articles deserve special thanks. I would also like to thank my father, wife and children who provided a favourable working atmosphere to me for successful completion of this project. Last but not the least, I would like to put a special word of appreciation, love and affection to Mohammad Imran Ali Khan who is a source of hope and inspiration.

NOOR MOHAMMAD

Department of Geography
Delhi School of Economics
University of Delhi
Delhi-110007

January 26, 1992

PUBLISHER'S NOTE

When Concept's International Series in Geography No. 1 — *Perspectives in Agricultural Geography* edited by Dr. Noor Mohammad in five volumes was issued over a decade back, we expressed our apprehensions that 'Posterity will be the best judge of our efforts in this direction. The response to this set of volumes will provide the impetus for hastening the publication of further series that are already on the anvil'.

Perspectives in Agricultural Geography received such an overwhelming response that it became out of print soon after its publication. Its success provided the necessary encouragement to issue thereafter Concept's International Series in Geography No. 2 — *Perspectives in Urban Geography* edited by Dr. C.S. Yadav in 16 volumes and No. 3 — *Perspectives in Geomorphology* by Dr. H.S. Sharma in four volumes.

We are glad to offer now Concept's International Series in Geography No. 4 — *New Dimensions in Agricultural Geography* edited by Dr. Noor Mohammad in eight volumes. In view of the new emerging trends, concept, theories and models in the field of agricultural geography, the set presents a wide panoramic view of the latest advances in this field. Hope this set too would be equally welcomed by the students, researchers, scholars and planners in various disciplines.

February 11, 1992 NAURANG RAI

CONTENT

LIST OF FIGURES

LIST OF TABLES

LIST OF CONTRIBUTORS

Bharara, L. P.
Social and Information Sciences Division, Central Arid Zone Research Institute, Jodhpur

Enyedi, Gyorgy
President, International Geographical Union Commission on Rural Development, Budapest, Hungary

Jain, C. K.
Department of Geography, Dr. Hari Singh Gour University, Sagar

Mohammad, Noor
Department of Geography, Delhi School of Economics, University of Delhi, Delhi

Morris, Arthur S.
Department of Geography, University of Glasgow Glasgow-G-12-800

Obara, Dunstan A.
Department of Geography, University of Nairobi, Nairobi, Kenya

Preston, Richard E.
Department of Geography, University of Waterloo, Waterloo, Ontario

Qureshi, Salahuddin
Department of Geography, Aligarh Muslim University, Aligarh

Ramachandran, R.
Department of Geography, Delhi School of Economics, University of Delhi, Delhi

Sharma, S. K.
Department of Geography, Dr. Harisingh Gour University, Sagar

Singh, Surendra
Department of Geography, J.V.M. College, Charkhi Dadri

Tewari, A. K.
Department of Geography, University of Jodhpur, Jodhpur

INTRODUCTION

DURING THE past three decades there has been tremendous methodological development in geography. One of the most striking features of these developments is the way in which overlapping areas with some of the behavioural services have grown. Behaviourial geography broadly falls into two major areas of concern : the first considers the behavioural aspects that are related to spatial patterns and movements, and explicitly recognises the effect of behavioural constraints in geographic models; the second recognises that human decisions with locational implications are affected by the way in which the decision-maker perceives the physical and human environment (Board, et. al., 1969).

The main function of this approach is to find process-oriented explanations concerning spatial phenomena rather than cross-sectional explanations that correlate elements of the physical structure of systems with human acts in such systems. The physical elements of existing and past spatial systems represent manifestations of decision-making behaviour on the landscape and they search for geographic understanding by examining the process that produce spatial phenomena themselves (Amedeo and Golledge, 1975). Geographic researches on spatial behaviour stem in large part form the fundamental paper of Wolpert (1964) in which linear programming and diffusion models are combined with stochastic sampling methods in a conceptual framework which recognises, for the first time in geographical studies, the non-optimal or satisfying aspects of spatial variations. Subsequently a number of research papers emerged dealing with various aspects of this major theme. Game theoretic model for the analysis of geographic patterns and movements (Dacey and Isard, 1962; Gould, 1963); spatial patterns and decisions modelled as learning process; travel behaviour at the micro-regional level within urbanised areas; effect of cultural variables upon spatial movements of rural people to lower order central place facilities (Riddell, 1965); geographic aspects of decision-making process (Wolpert, 1965);

transportation development as searching process (Gould, 1966) are some of the major areas of research. However, the basic thrust of research in spatial behaviour has the objective to discover those shared characteristics which might provide clues in developing methodological and theoretical insight through analysis of the perception, evaluation, learning and response of individual to their environment (English and Mayfield, 1972).

The second major trend in behavioural geography (Environmental Perception) forms an overlapping area with certain aspects of the field of psychology (Lowenthal, 1961, 1967). In some cases either the geographers have training in geography and psychology both or have worked in collaboration with psychologists of similar interests. Nonetheless geographers approach to concept of environmental perception in a variety of ways : some are quite simple, while others are complex. At simplest level, environmental perception assumes that each individual has an image of the world, and that his preferences, evaluations, decisions and subsequent behaviour are based on the world of objective reality. Given this premise, man's behaviour with respect to environment becomes the central theme of the research problem. Those forces which contribute to environmental and spatial behaviour such as past experience, available information, physiological conditions, philosophical traditions, religious taboos, social constraints and communication spheres become the legitimate concern of the geographer. In other words, once human geographer seriously studies human and their spatial perceptions and behaviour, the entire range of problems which baffles other social scientists become their concern as well, and it is here that the complexity begins and problems emerge. (English and Mayfield, 1972).

However, human geographers have developed three distinct approaches in this connection. Peter Gould's 'Mental Map' represents one of the behavioural approaches to the definition and analysis of perceptual environment, although it departs from earlier writing in behaviour literature which assumed a rather narrow normative model of economic man in which each decision-maker was blessed with perfect knowledge, power of prediction and economic rationality (Wolpert, 1964). The main objective is to devise a technique to measure the ways in which people perceive and evaluate geographical space in terms of residential desirability. These perceptions are critical in the formation of those decisions which restructure the human landscape, and these decisions vary with the experience, knowledge and aspirations of each individual.

The second approach relates the concepts in behavioural sciences to the perceptions and utilisation of resources. It deals with the quality of environment, specially in terms of the ways in which resource managers and users respond to natural hazards. The search is for behavioural regularities which might contribute to theory building and to intelligent policy decisions with respect to resource utilisation. Several studies have been conducted to integrate intensive analysis of components in the physical environment with detailed examination of human attitudes and responses to stimuli such hazards (Burton, 1961; Burton and Kates, 1964; Kates, 1963; Ramachandran, 1973; and Saurinen, 1966).

The third approach to integrate spatial behaviour is the direct analysis of the decision-making process. Decisions and their spatial expressions are influenced by the perception of the environment and the information available to the decision-maker. The environment is not derived simply from the elements of the physical world, but includes the elements of cultural, social and political setting (Abler, 1971). However, various decision models developed in the social sciences have been applied to locational decision-making, viewing the set of choice among existing alternatives not only within an economic framework, but also in terms of knowledge, attitudes and anticipations of the decision-makers. However, there are three major areas in decision process which have attracted more attention than others : (1) decision objectives and goals particularly in the location of public facilities in planning; (2) the degree of rationality in the decision process as gauged by the concepts of adaptive rational behaviour and bounded rationality; and (3) the impact of stress and threat as environmental factors present in the decision environment. The most important underlying principle of environmental perception is that man's decisions and actions concerning his environment are based not only on objective factors but also on subjective ones. Several methods have recently been developed to study man's interaction with various types of ecosystems, but many physical and social scientists are unfamiliar with the possibilities for systematic observation in the field of subjective perception. Hence, there is a need for methodological studies which would both provide a rationale and description of the field of environmental perception in the context of man biosphere relations and ecosystem management. **Obara's** paper is an attempt in this direction in which he has examined the applicability and suitability of repertory-grid technique for investigating and interpreting environmental perception and behaviour with special reference to small scale farmers in developing countries. It is a comparatively new technique which enables the researcher to find out as to what things are put in the same or different categories in people's minds and how the classes of phenomena are distinguished from each other. It is designed to study the perceived world of an individual with minimum of interview interference (Toronstend, 1977). However, the author after listing the common characteristics of several forms of repertory grid methods, has discussed its use in geographical research. He has given the repertory-grid design with empirical example, and has explained the techniques for analysing repertory-grid data. Finally, the author has discussed the merits and limitations of repertory-grid system.

Droughts and floods are the two major vagaries of Indian agriculture. Both the events occur in one part of the country or the other. The government is injecting large sum of money in combating with these problems specially in case of arid and semi-arid regions, without, perhaps, taking the advantage of long experience of the farmers who can mitigate this impact of calamity in their own way. This inability has brought frequent miseries and successive famines to the arid land farmers whose economy is totally shattered during and after such events. This is more so evident in the desert fringe villages where despite great physical susceptibility to rain failure, the rural density of population is high. Perception is a four stage psychological process of human behaviour where an individual senses a situation in his own perspective, thinks of it, remembers it, and finally reacts to it. Accuracy of perception makes an individual more satisfied and better adjusted to hazardous situation. **Tewari** in his paper has examined as to how far this behavioural process could be made useful in the adaption of and adjustment to environmental risks in Luni basin lying in south-western margin of Great Indian desert. The author has analysed the inhabitants impressions obtained by a questionnaire survey of 751 households in 5 selected villages representing the three geo-economic gradients of a desert region viz. arid, semi-arid and sub-humid. Spatial variations in hazard perception have been analysed in terms of awareness of drought severity, the frequency of occurrence, variation in the degree of personal experience and the type of farm operation. The central theme is to evaluate the farmers conceptualisation of the problem and their instant connection to uncertain events. The study reveals that long term residents, more experienced farmers, and those whose economy is more vulnerable to the variations of the weather, perceive the risk more realistically and are better adapted. In order to alleviate the problems of drought with a new perspective, there is a need of better appreciation of farmer's conceptualisation of droughts and native measures of diluting its impact.

Bharara in his paper has also assessed and analysed the perception of drought and desertification by the inhabitants of the Indian desert including 11 districts of western Rajasthan. The study is mainly based on primary data collected from 309 sample respondents chosen from 12 villages on a multi-stage stratified purposive sampling design in which the region was divided into three ecological zones viz. pastoral, rainfed and irrigated, and from each zone a district and, from a district 4 villages were selected. The respondents were selected from different religious, castes, occupational structure, and size of holdings. Secondary data are also used for general geographical character of the region and historical evidences and records. The author after explaining the concept of drought perception, examines people's perception of drought hazards which not only varies in different ecological zones but also among different socio-economic groups, and suggests a strong positive relationship with aridity which is also proved when analysed in terms of *Zamana* and frequency of occurrence of drought. The respondents have reported a large number of problems of socio-cultural, economic and bio-physical in nature which also vary in different ecological zones. It results in desertification and the nature and extent of parametres of perceived desertification such as climatological, ecological, socio-economic, and cultural, have been studied in depth over a period of 30 years. The recollection of drought by the respondents of Jaisalmer, Pali and Nagaur over a period of 93 years is also analysed and a correlation of rainfall records and, recollection of *Zamana* is worked out which shows very positive relationship. In fact the recollection of drought is associated with intense storms, migration, shifting dunes, livestock mortality etc. However, the study of this type of thinking may lead to a letter understanding of non-scientific thought but it will serve a key to understanding the coping mechanisms and potential reactions of a rural population.

Diffusion studies have become very common during the third quarter of the twentieth century among the scholars of the West. This approach is particularly applicable to the study of growth for single innovation for short period of time. Its applicability to more complex growth process is less clear. Hence, it is criticised as a general model for spread process because it focuses on spread of phenomena without modification of the object being spread in its environment. **Morris** in his paper has critically examined the growth process of the *colonia agricola* i.e. organised farming colonies in nineteenth century in Argentina. The author has firstly critically examined the concept of diffusion, deformation and epigenesis with the view as to how far these concepts are applicable to the process of growth of *colonia agricola* in Argentina. These spread around centres in southern and central Santa Fe, but changed in their essential character over the diffusion period 1850-1900. But these changes can be scarcely referred to a diffusion with deformation, as might first be supposed. They represent more complicated changes in structure with growth, and this joint process is termed epigenesis. The analytical review of the growth process of *colonia agricola* indicates at least three levels of growth process, i.e. major systems growth process; growth process of institution and complex concepts; and growth process of individual innovations which are closely interlinked.

The soul of green revolution is the miracle seed known as high yielding variety (HYV) which has spatio-temporal variations in its diffusion. Among a host of factors, environmental factors specially climate plays decisive role in adoption of HYV in Madhya Pradesh as is revealed in the study of **Sharma**. The author, after giving general trend and spatial pattern of adoption of HYV of wheat, rice, jowar, maize and bajra in the state, analyses its use in Sagar district and Bhopal, Vidisha plateau on the basis of 100 and 143 selected samples respectively. The study examines the relationship between climate and HYV of rice which shows that on the whole the climatic condition is not very conducive for its adoption, leaving the exceptions a part. Hence, the farmers prefer the improved indigenous varieties. Wheat has shown a contrary result and it does not face the environmental problem. Generally, it is grown with irrigation which is available. It helps not only in adoption of HYV of wheat but also use of chemical fertilisers and plant protection chemicals. Jowar — the third ranking foodgrain in the state faces the problems like rice and the area under HYV of jowar is not very encouraging. However, the study concludes that since, the climatic milieu can not be changed, improved strains suitable to it must be evolved locally rather than importing them from foreign climate. Secondly, attention has to be paid towards water management, which will save the plant from destruction at the time of excessive rains as well as in drought.

As a result of introduction of various farm innovations in the past three decades, there has been considerable change in the farming methods and practices all over India. But the rate of adoption of these innovations vary from region to region and even from community to community. These differences can ultimately be traced to the level of individual farmers and his characteristics. **Ramachandran** explores the differences in response of individual farmers to innovations in the Ganga Plain. His study is based on field interviews of 480 randomly selected heads of the farm household belonging to Jat, Rajput, Gujjar and Muslim communities in 12 villages of Meerut district. Each farmer was interviewed regarding the adoption of a set of 64 innovations. The author has worked out the innovations score with the help of a statistical formula and has critically examined the association between the innovativeness and social groups, location with reference to urban centre, personal characteristics of the farmers, farm characteristics, exposure to mass media, cinema and documentary films and daily newspapers and magazines. The study reveals that average innovativeness scores vary remarkably among the 4 social groups. The Jats and Rajput farmers have high average scores, while Gujjar and Muslims have low innovative scores. The study further reveals that farmers' personal characteristics such as education, age, family type, family size, perceived social status of the farmers are also related to innovativeness. However, these relationships are not simple but complicated.

Diffusion has been the subject of geographic enquiry for several decades. Spread of phenomena, idea, technique or thought throughout a population or region incorporate the basic geographic elements of distance, direction and special variations, and this forms a valid field of geographic interest. A review of diffusion studies indicate that it is characterised by two distinct approaches (Board, et al., 1969). The first approach is generated within the context of cultural geography whose research goal is to identify environmental features, characteristics of a given culture, and, if possible, to discover what role human action has played in creating and maintaining given geographic features. It distinguishes, describes and clarifies the typical complex of environmental features including the man-made ones that coincides with each cultural community (Wangner and Mikesell, 1962). In the second approach attention turns to the identification of generative process by which the observed locational pattern of a given phenomena comes into existence (Sauer, 1952).

However, the pattern of diffusion of innovation is not uniform both in time and space. It is influenced by physical, social, psychological, economic and demographic factors. Although various studies have been done abroad to highlight the impact of these factors on diffusion of innovations, there is still lack of geographical studies in this field in India. **Mohammad** in his paper has made an attempt to ascertain the impact of socio-demographic factors on diffusion of agricultural innovations in Saryupar plain of northeastern Uttar Pradesh. The author has selected four major groups of innovations viz. chemical fertilisers, improved varieties of seed, improved agricultural implements and plant protection measures, and has assessed the impact of seven socio-demographic variables i.e. education, training, age, caste, value-orientation, occupation and size of family on adoption of above innovations. The farmers constitute the ultimate unit of study and a total of 1249 articulate farmers are selected from 45 sample villages lying in different homogeneous strata of the region. The study is based on primary data collected by the author in 1974-75 at three levels i.e. block, village and farmer. The data are processed and each respondent is assessed to different parametres as well as adoption rate according to scoring schemes evolved by the author for quantifying the variables. Hypotheses are formulated regarding various factors and analysis of variance and 't' tests have been used in order to test these hypotheses. The study reveals that socio-demographic factors play very significant role in the diffusion of agricultural innovations. In the conclusion various measures have been suggested to improve the economic condition of the farmers.

No doubt the heavy induction of modern technology has brought a revolution in agriculture but it is not uniform in time and space due to variations in environmental, techno-economic and socio-institutional conditions. **Jain** has described and explained the trend and pattern of adoption of modern farm technology such as irrigation — its sources, area irrigated, and its spatial pattern; high yield varieties of seed — rice, wheat and maize; chemical fertilisers — types and spatial distribution; plant protection measures and farm implements — iron plough, diesel oil pump, electric pumping set, tubewells, tractors. On the basis of degree of adoption of these 6 innovations, the author has worked out composite index of level of adoption by giving rational weightage to each innovation and have grouped the districts into six levels of adoption — very high, high, moderate, low, very low and extremely low — which have shown a definite pattern of distribution. The study reveals a depressingly low degree of adoption of modern farm technology. There is wide regional variation in the level of adoption. Western and northwestern districts have high and very high level, while the eastern half of the state is far behind the state level due to marginal and small holding, high percentage of tribal and scheduled caste population, high pressure of rural population, low level of literacy, and lack of input and credit facilities.

The speedy and extensive development of agriculture, by and large, depends on technological change and spatial diffusion of agricultural innovations. Since the World War II vigorous attempts are being made for mechanisation of traditional agriculture in Africa. The use of tractor has frequently been considered general panacea for the problem of increasing productivity in East Africa. Several other costly programmes involving purchase of large quantities of equipments and a crash training programmes for the large scale mechanisation were practised with inadequate preparation in sixties but the running costs were too high to be compensated with increase in production. Despite these problems, schemes have been streamlined to manageable levels. The main objective being to assist farmers to achieve timely cultivation and to bring more land under cultivation. But this partial mechanisation is likely to aggravate labour bottleneck later in the season, when operations are conducted manually. To obtain an efficient sequence of operation, mechanisation must alleviate and not accentuate labour shortage in the small scale farming system. Where the available farm labour is such that it limits the economic expansion of production, farmers may find it feasible to substitute capital in terms of mechanisation. If mechanisation is considered as substitute of mechanical aids for mechanical and human power, it is significant to note that small holder cotton production in East Africa is still lagging behind a progressive mechanisation. **Obara** in his paper has highlighted the problems and opportunities of small holder agricultural mechanisation. He has examined the determinants of farm mechanisation in East Africa such as size of holding, fragmentation of farms, employment problem, financing, investment uncertainty, yield levels and environmental factors. Finally he has come to meaningful conclusions and has recommended practical solutions to all these problems.

Since Independence several agricultural innovations have been introduced in India in order to increase agricultural production to become self-sufficient. It has brought tremendous change in agricultural practices and has increased agricultural production substantially. But the pattern of diffusion of these innovations is not uniform both in space and in time, perhaps, because of the varied physical, social, economic and technological conditions. There are states like Punjab and Haryana where these innovations have considerably been adopted, while in the states of Orissa and Madhya Pradesh the adoption rates have been very slow. A few studies have been made to ascertain the causes of differential adoption rates, but they suffer from over generalisation. Hence, **Mohammad** in his paper has made an attempt to examine the impact of economic factors on the diffusion of agricultural innovations at micro level in Trans-Ghaghara Plain consisting of the districts of Bahraich, Gonda, Basti, Gorakhpur and Deoria of northeastern Uttar Pradesh. Four major groups of agricultural innovations, viz., chemical fertilisers, high-yielding variety of seeds, improved agricultural implements and plant protection measures have been selected. The author has assessed the influence of nine economic variables i.e. the size of the holding, land tenure, irrigation facilities, power and diesel, credit, subsidy, input, yield and fair prices. The region has been divided into five homogeneous strata on the basis of physical, social and economic conditions, and 45 villages have been selected from different strata on the basis of communication status. The farmers constitute the ultimate unit of study and a total of 4,249 articulate farmers are selected. The study is based on a primary data collected by the author in 1974-75 at three levels i.e. district, block and village. The data are processed and each respondent is assessed to different parametres as well as adoption rate according to scoring scheme evolved by the author for quantifying the variables. Hypotheses are formulated regarding each variable and analysis of variance and 't' tests have been used in order to test these hypotheses. The study reveals that all the selected variables play very significant role in the diffusion of agricultural innovations. In the conclusion various measures have been suggested to accelerate the rate of adoption of agricultural innovations and consequently to improve the economic condition of the farmers.

Heavy induction of modern farm technology in Indian agriculture, termed as Green Revolution, has a number of socio-economic implications on the rural landscape. On the one hand, it has changed the traditional subsistence farming system to market-oriented production system, on the other hand, it has resulted in multi-facet socio-economic improvements in rural areas. **Singh** in his paper after discussing the inherent concept of Green Revolution and its acceptance in global perspectives, has analysed the genesis of Green Revolution and the specific reasons for selection of the areas of its introduction in India. The author has accounted the achievements of not only Green Revolution, but also the various infrastructural developments which are ultimately targeted for increasing agricultural production leading to comparatively more properly and socio-economic transformation. It has averted famine and hunger despite the burnt of population explosion. It has positively affected landholding size, position and status of labour — both nature and immigrant, marketing system and so on. However, it has resulted into several ecological problems such as land and soil degradation, contamination of water and food depletion in forest, and groundwater resources. The study concludes that for development of India, rural development is inevitable and the Green Revolution has provided the required thrust. So it should be extended to peripheral and distant suitable areas for creating balanced development but not with the widening economic gap and socio-cultural problems. The co-operative movement needs to be further strengthened to create job opportunities in rural sector. Besides, rural industrialisation and patronisation of their products need fresh assessment and planning.

Every technology acts as a vehicle of change and modern farm technology is no exception to this rule. Leaving exceptions apart, improved farm technology has acted as unfailing engine to accelerate the pace of socio-economic development in our country. Compared with the general poverty of pre-Independence period, the farmers of today have undoubtedly reached a stage of relative prosperity and well-being. Their standard of living, housing condition, agricultural infrastructure, household assets, degree of mobility, attitude towards education and profession of their children, religion, culture etc. have changed and improved. In fact the new found prosperity has brought about a complete transformation in the socio-economic life of the countryside.

But the diffusion of farm technology is not uniform all over the country both in space and time, and consequently, socio-economic consequences also vary considerably. **Mohammad** has studied this issue in two geo-economically contrasting regions — one is the predominantly fertile agriculture tract — Ganges basin, and the other is mountainous region — Himalayan region. The study is based on primary data collected from 300 respondents belonging to 5 selected villages of Etawah district (U.P.) and 6 selected villages of Baramula district (J & K). The respondents have been interviewed according to seven exhaustive survey schedules including geographical setting of the village, personal and households characteristics of the respondents, adoption of agricultural innovations, mass-media exposure, social and economic implications of technological change in agriculture. Eight socio-economic indicators viz. education, size of family, political awareness and involvement, mobility, occupational structure, per capita income, housing condition and media exposure, have been examined in this study.

The study conducted in two distinct regions of northern India comes up with several interesting results: some of them are entirely new, while a few are in conformity with the existing ones in the field. It has broadly been found that the level of technological change in agriculture is very much associated with socio-economic level of respondents. There is considerable inter- and intra-regional disparity in the level of innovativeness among the respondents. Generally the respondents of progressive villages are more receptive to new farm technology leading to their better prosperity as compared to the respondents of non-progressive villages. Similarly, the respondent of the U.P. villages are, by and large, more innovative as compared to their counterparts in the village of J & K. Same is true with regard to the socio-economic implications. However, of the eight hypotheses proposed by the author six stood the test of empirical analysis and were approved with a fairly high degree of significance. Of the remaining two, one is disapproved, while the verdict in case of other is turned out to be tentative. The innovative farmers were found having a clear edge over their traditional counterpart in terms of education, political involvement, mobility, occupation, housing, income and media exposure.

The study concludes that modern farm technology and socio-economic development are entwined in a mutually supportive relationship. The two together form an interesting cycles where each is a means of achieving the other. The message is crystal clear. Advanced farm technology undoubtedly leads to socio-economic well-being and it in turn sets the farm technology on wheels towards a greater well-being, and so the cycle remains in motion.

Thus the devil of rural backwardness could be banquished by

adopting the strategy consisting of a two pronged attack i.e., large scale induction of modern agricultural technology, and simultaneous efforts towards creation of socio-economic climate congenial to it. The former is possible when an efficient agricultural infrastructure is developed, while the later could be created by raising the level of education, maximising the exposure of farmers to mass media, and revitalising the near defunct extension services.

In an other article **Qureshi** has screened the severe economic stresses inflicted by Green Revolution on the survival mechanism of subsistence farming in the Ganganagar district of arid western Rajasthan. The author emphasises that the benefits of Green Revolution which comes as subsidy for various agricultural inputs and other incentives for development of agricultural infrastructure provided by Resource Arid Scheme are more tangible for the already established large farmers. The SFDA and MFALA could not fetch relief due to the inbuilt weaknesses in the mode of relief. As a result commercial and market-oriented agriculture has developed and the marginal farmers sell their economically unfeasible land to big farmers due to input deficiency. However, the study concludes that green revolution along with its production surplus and resultant prosperity to commercial farmer has made survival of subsistence farmers more difficult than before. The economic conditions of the subsistence dryland farmers in Ganganagar tehsil is miserable because subsistence farmers have only 11 per cent of the total cultivated land against 89 per cent irrigated land. Hence, development and protection priority should be given to the areas which are facing more intense competition of survival where subsistence farmers are in growing economic discomfort.

In the development of a region, marketing facilities play a significant role as they are points of transaction of agricultural input and output. Small urban places in rural areas play a vital role in this connection. **Preston** in his paper, although the main focus is not on marketing of agricultural input and output, has dealt with the central place function and growth centre potential of small urban places in rural areas and their contribution in rural development. The study is related to the Pacific northeastern section of United States including the states of Washington, Oregon, Idaho and mountainous counties of western Montana. This is an area with strong and varied physical geography. After discussing in brief the population distribution and important urban centres, he has used Christaller's model with modification in order to determine the central places.

The earlier studies have examined the various facets of agricultural modernisation and its socio-economic implications but **Enyedi** in his paper has assessed its impact on rural settlements in Hungary where a lot of inter-sectoral fluctuations in economy have taken place. The author has identified several phases of agricultural development such as phase of mechanisation and economic concentration, phase of biological revolution, phase of modernisation of economic organisations and so on, and has given a detailed account of each phase. Although the author has accounted briefly the general characteristics of the transformation of the rural sphere during the past thirty years (size of rural population has decreased minimally, socio-economic structure of village and city have become similar to each other, predominant agrarian character of the rural sphere has become multi-functional, and modernisation of rural functions has been faster than modernisation of rural settlement infrastructure) he has examined it in detail with reference to (i) Internal transformation of agricultural population itself; (ii) New ties between large-scale farms and rural settlements; and (iii) Development of regional ties within rural sphere. However, the study concludes that rural settlement had been harmonious with the surrounding agriculture the village, the topographical location of its population, its internal structure, its houses, the character of its services, the road network were all adapted to the needs of agriculture itself. The socio-economic development of the past 30 years had broken that traditional harmony once and for all. As a result of the rapid social transformation, the rural settlements must serve more than the needs of agriculture only. But the demands of agriculture have also been altered significantly as a result of the technological development and the agro-industrial integration during the last decade. The technical infrastructural development of the rural sphere was considerably behind that of agriculture. In fact, it was not even able to satisfy the needs of the latter. As a result tension has develop between the village and the agriculture, the resolution of which is necessary both because of the needs for increased agricultural output and because of the need to improve the standard of living of the rural population. The rural settlement network attempts to adapt quite spontaneously to the changed social environment for which it provides a framework. It is high time to attempt to define the rural model and the rural-urban relationship of the developed socialist system in order to allow our settlement planners to plan for the entire settlement network and not to restrict its development merely to our cities and urban spheres.

REFERENCES

Abler, R., Adams, G. S. and Gould, P., 1971 : *Spatial Organisation : The Geographer' View of the World*, p. 515.

Amedeo, D. and Golledge, R. G., 1975 : *An Introduction to Scientific Reasoning i. Geography*, p. 348.

Board, C. et al., 1969 : *Progress in Geography : International Reviews of Curren. Researches*, p. 22.

Burton, I., 1961 : *Changes in the Urban Occupance of Flood Plains in the United States*, Research Paper No. 57, University of Chicago.

Burton I. and Kates, R. W., 1964 : "The Flood Plain and the Seashore : A Comparative Analysis of, Hazard-Zone Occupance", *Geographical Review*, Vol. 54, pp. 366-85.

Dacey, M. F. and Isard, W., 1962 : "On Projection of Individual Behaviour in Regional Analysis", *Journal of Regional Science*, Vol. IV, pp. 1-34.

English, P. W. and Mayfield, R. C., 1972 : *Man, Space and Environment: Concepts in Contemporary Human Geography*.

Gould, P. R., 1963 : "Man Against His Environment: A Game Theoretic Framework", *Annals of the Association of American Geographers*, Vol. 53, pp. 290-97.

———, 1966 : "Space Searching Procedures in Geography and the Social Sciences", *University of Hawaii, Papers in the Social Science Research Institute*, 1.

Kates, R. W., 1962 : *Hazards and Choice Perception in Flood Plain Management*, Research Paper No. 78, University of Chicago.

Lowenthal, D., 1961 : "Geography, Experience and Imagination Towards a Geographical epistemology", *Annals of the Association of American Geographer*, Vol. 51, pp. 241-60.

———, 1967 : *An Analysis of Environmental Perception*, New York.

Ramachandran, R., 1973 : "Human Perception of Flood Hazards in the Ganga Plain", in White, G. F. et al., *Natural Hazard Perception and Choice*.

Riddell, B., 1965 : "Statistical Methods in Geographical Research", *Geographical Review*, Vol. 46, pp. 129-32.

Saurinen, T. F., 1966 : *The Perception of Drought Hazard on the Great Plains*, Research Paper No. 106, University of Chicago.

Wolpert, J., 1964 : "The Decision Process in Spatial Context", *Annals of the Association of American Geographers*, Vol. 54, pp. 537-58.

———, 1965 : "Behavioural Aspects of the Decision to Migrate", *Regional Science Association : Paper and Proceedings*, 15, pp. 159-69.

[illegible]

[illegible]

[illegible]

[illegible] and [illegible] [illegible]

[illegible] *[illegible]* [illegible]

[illegible] *[illegible] Research* [illegible]

[illegible] *[illegible]* [illegible]

[illegible] *Journal of [illegible] Science*, Vol. [illegible]

[illegible] [illegible]

[illegible] *[illegible]* [illegible]

[illegible] [illegible]

[illegible] *[illegible]* [illegible]

[illegible] [illegible]

[illegible] [illegible]

[illegible] [illegible]

[illegible] [illegible]

1

ENVIRONMENTAL PERCEPTION AND BEHAVIOURAL APPROACH IN AGRICULTURAL GEOGRAPHY

DUNSTAN A. OBARA

MAN'S DECISIONS and actions concerning his environment are based not only on objective factors, but also on subjective ones. Perhaps, this is the most important underlying principle of environmental perception research. But attempts to promote environmental perception research have been hampered by a lack of awareness of available methods which have recently been developed. This constraint has, therefore, affected the incorporation of a perception approach into the field studies dealing with man's interaction with various types of ecosystems (for example, agricultural systems) and oriented towards the solution of concrete problems. Since many physical and social scientists are unfamiliar with the possibilities for systematic observation in the field of subjective perception, it has become increasingly apparent that there is urgent need for methodological studies which would both provide a rationale and description of the field of environmental perception in the context of man-biosphere relations and ecosystem management. In the light of the foregoing problem, it is the main objective of this paper to examine a methodology suitable to investigating and interpreting environmental perception and behaviour with special reference to small-scale farmers in developing countries. Some of these farmers are in transition from subsistence agriculture to commercial farming, thus they are often faced with a multitude of agricultural problems. The farm problems may be solved through the application of repertory-grid technique, although it has its limitations.

Theoretical Basis of Repertory-Grid Techniques

Several disciplines — anthropology, geography, architecture, psychology, sociology, urban and regional planning — have applied the environmental perception approach to some extent. However, in a number of disciplines, the significance of individual and cultural perception of the environment for human behaviour began to be discussed in the 1950s. As a multi-disciplinary field with some common research problems, environmental perception study was mainly established in the early 1960s. Before this latter period, there was little sense of impetus or cross-fertilisation of ideas on the subject. Consequently, during the last decade, environmental perception has become a rapidly expanding field of research interest that is empirically and practically oriented. The repertory-grid technique has been specially developed to facilitate research in environmental perception and behaviour. To date repertory-grids are devised to elicit cognitive systems, differing from other quantitative approaches such as questionnaires in that the individual or the group is expected to generate a personal set of questions and answers. The technique is, therefore, used to determine how people categorise phenomena : droughts, floods, pests, diseases, soil erosion, places, incomes and other events. Moreover, continuing application in psychiatry has yielded wide experience and understanding of its mathematics of analysis (Slater, 1974). It enables the researcher to find out what things are put in the same or different categories in the people's minds and how the classes of phenomena are distinguished from each other. The technique is designed to study the perceived world of an individual with minimum of interview-interference (Townsend, 1977). As part of his broader theoretical framework of Kelly (1955) introduced the central idea of the repertory-grid technique as a clinical technique of measuring personality between people in their personal lines. Since its introduction, the technique has been used further (Oppenheim, 1966; Bannister and Mair, 1968) to measure, among other things, tribal perception of urban services, and comparisons of different places — locally and nationally. As used in environmental perception, its results are not necessarily interpreted within the scope of Kelly's Personal Construct Theory (PCT) from which it originated as a model of personality. To date, several other forms of repertory-grid have been developed with specific details (Bannister and Mair, 1968), although they all have certain characteristics in common (Bannister and

Frensella, 1974) :

(1) They are concerned with eliciting the relationships for a person between sets of constructs, either in terms of construing elements or by directly comparing construct with construct.

(2) The central aim is to reveal the construct patterning for a person and not to relate this patterning to some established normative data. There is no reason why normative data should not be collected for a specific purpose as in the Grid Test of Thought Disorder, but the individual construing system is the prime focus and the standard test forms an occasionally useful venture.

(3) There is no fixed form or content. It is called repertory-grid technique and not test, and the selection of the form and content is related to each particular problem. A grid designed to investigate how nomadic bushman interprets his environment would be meaningless to a suburban commuter, even in translation, except perhaps to show that the commuter might have problems if he had to travel in the terrain of the bushman.

(4) All forms are designed so that statistical tests of significance can be applied to the set of comparisons each individual has made. A basic assumption underlying the method is that the psychological relationships between any two constructs, for a given person, are reflected in statistical association between them when they are used as judgmental categories. The foregoing theory is confirmed by Oppenheim (1966) who pointed out that from each variant of the technique we can usefully infer the way a person "sees" the world from patterns of associations between constructs and objects. This confirmation is further reinforced by the fact that repertory-grid is a technique devised to establish how an individual sees the part of the world in which the researcher is interested in his own terms (Dixon and Leach, 1978).

The repertory-grid technique is not a new field in geography since it has largely been used by behavioural geographers (Harrison and Sarre, 1971, 1975; Hudson, 1974). Nevertheless, practically any individual can be administered a grid provided that a suitable form is constructed. (Chetwynd, 1974): grids are basically geared to communication problems (Bannister and Mair, 1968) and may help geographers extend their perception studies beyond cultural boundaries (Townsend, 1976). Baker et al (1977) demonstrated that traditional farmers were a potentially rich source of ecological information, supplementary to that retrieved from more expensive and sophisticated scientific sources. Moreover, their interpretations of ecological changes are sound. Similar studies have been carried out in West Africa (Oguntoyinbo and Richards, 1978). These studies have revealed that farmers had no difficulty in understanding and performing the basic test. The farmers were able to identify dangerous pests from field specimens and report the damage they did and many were aware of the seasonal relationships between their appearance and rainfall, their affinity for the pernicious weeds and the way the pests tended to behave in the field. However, developments over the last two decades have been so complex that there is some debate whether the technique can be considered generalised and free from theoretical assumptions as maintained by Chetwynd (1974) or whether it must be based on Kelly's (1955) original PCT, which was considered as seminal work. But the way in which constructs are defined is both empirical and operational (Lundeen, 1972).

Repertory-Grid Design

A repertory-grid or matrix represents the respondent's system of personal constructs, in which the columns of the matrix consist of the elements and the rows contain the discriminators or constructs that are used to classify the elements (Table 1.1). In the context of the traditional farmer's environmental perception, the elements might be pests, weeds, diseases, droughts, floods, or farm problems in general and could be provided by the respondent, the interviewer or by a combination of both. Although the grids represented to the respondents may be similar to those of semantic differential scales, the constructs used would be in some sense his own and not those devised by the researcher. In a survey of farm problems it seems more valuable to allow the farmer to design his problems, either during the interview or from an earlier questionnaire survey in which he participated. Alternatively, if the elements were pests, it would be useful to include automatically those of direct research interest and let the farmer add others which he thought were of paramount importance.

Whyte (1977) provided useful information on the repertory-grid technique. The following presentation of triads is partly based on his work and partly on the author's example. The technique is one of presenting a respondent or group of respondents with triads, or sets of three phenomena or concepts. These are often words either written on cards or spoken verbally, but they can also be pictures on cards, drawn on the ground or pointed out in the environment. Cards have the advantage that the respondent can handle them, and physically rearrange them as he considers alternative groupings so that the task becomes simpler, more concrete and evidently more enjoyable. For example, mountain, river, field, village and the like could be presented in several different ways depending on the interview conditions. These are called elements, which are shown on the columns. They may either be given as a set to the respondent or he may produce his own as indicated above. Often responses to open questions in the rest of the interview can provide the set of elements, although the element set can alternatively consist of some researcher-initiated responses and some respondent-initiated ones. The respondent is then asked to show one important way in which any two elements of the three presented to him are alike and are different from the third element, for example, he might say mountain and field are alike because both provide him with food. Then he is asked which card is the odd one and in what way it is different from the others. In this case if river were the third element in the triad, he might say it was different because it did not provide food but flooded his crops and destroyed them. To another respondent one could imagine a quite different set of constructs and groupings for these same three elements, reflecting differences in environmental perceptions and lifestyles. For each of the remaining cards in turn the respondent is then asked to say whether they are more on the "alike" side or the "unalike" side of the construct he has defined. Alternatively, the respondents may be asked to rate the elements according to how much alike or unalike they are on the construct (Tables 1.1 and 1.2), but this is a more difficult task for a respondent in an already long test and requires more analysis.

The elements are randomly presented in sets of three, ideally until all possible combinations have been explored. This full repertory-grid permits analysis of the relationship between respondents' constructs by factor analysis or principal components analysis in order to identify the major principles on which he discriminates between phenomena. But the triad method of presentation is being increasingly used in field research independently of repertory-grid techniques. For simplicity the interviewer's task is to show a grid and a set of typical instructions to interviewers (Table 1.1).

The following instructions provided by Whyte should be used in completing the above repertory-grid :

1. Interviewer places three cards selected at random in front of the respondent and asks him to say one important way (construct) in which any two of them are alike and are different from the third one. Write his response in the "alike" column.
2. Then ask which is the odd card out and in what way it is different from the others. Write down his response in the "different" column.
3. Write X in the appropriate boxes for the two "alike" elements. Write O in the appropriate boxes for the "different" elements.
4. For each of the remaining cards in turn ask him to decide whether it is more on the "alike" side or the "different" side of the construct. Write A (alike) or D (different) for each box in the row as he answers.
5. Repeat for as long as possible, or until the grid is completed.

A further example comes from the author's research in 1979. Small-scale farmers were shown five different pests, which they were then asked to place in groups of three and state in what ways two were similar and different in terms of their cotton farming problems in the Kano Plain of Kenya. Table 1.2 shows five elements (pests) and constructs. In this last example, the scoring procedure adopted is simply presence or absence (yes or no), although more sophisticated ranking and numerical techniques have been developed (Bennister and Mair, 1968). After considering the first three elements, the next three elements from the total set are presented to the farmer and he is asked to repeat the procedure. Possibly some of the same constructs are cited, together with new ones and these are added to the grid. The interview continues until all combinations of elements have been exhausted, and the repertory-grid for that individual has been completed.

Different people produce different constructs for each combination of elements, and probably a different total number of constructs may be expected. If the survey requires that farmers give their own set of elements at the initial stage of the interview, they are also likely to produce different total numbers of elements. In the last example, the elements were cotton farm pests, but an equally valid strategy might have been to make the elements farm crops or farm problems in general.

TABLE 1.1: Repertory-grid survey

Constructs Alike	Hill 1	Road 2	Village 3	Well 4	Cornfield 5	Stream 6	Forest 7	Bean field 8	Farm 9	Pasture 10	Different
1. Dry	X	A	A	D	D	O	D	X	D	A	Wet
2. Home	A	D	X	A	X	A	O	A	A	A	dangerous
3. Ours	D	O	A	X	A	D	D	A	X	D	not ours
4.											
5.											
6.											
7.											
8.											
9.											
10.											

Note: The Elements and Constructs filled in are samples only.

Source: Whyte, A.V.T. (1977): *Guidelines for Field Studies in Envirormental Perception*, UNESCO.

In summary, it is said that repertory-grids form a flexible method of evaluating and eliciting mental constructs, which leave much to the inventiveness of the researcher with respect to the problem in hand (Baker, 1977). This is succinctly summarised by Hudson (1974) as: A flexible method of allowing people to structure their images of reality in their own terms (within the assumptions of the method itself) rather than forcing people to reply in terms of concepts supplied by the researcher.

TABLE 1.2: **Repertory-grid from an individual farmer**

	Elements					
	Rank	Pink Bollworm	Spiny Bollworm	American Bollworm	Lygus	Stainer
Boll-shedding—no boll shedding	1	1	0	1	0	0
Spoil boll—does not spoil boll	2	1	0	1	0	1
Persistent—non-persistent	3	1	0	1	0	1
Resistant to DDT/Servin—not resistant to DDT/Servin	4	0	0	0	0	1
Shoot destroyed—no shoot destroyed	5	0	1	0	0	0
Punched leaves—not punched leaves	6	0	0	0	1	0
Stays under leaves—not found under leaves	7	1	1	0	0	0
Weeds encourage—no encouragement by weeds	8	0	1	1	0	0
Eaten by lady bird—not eaten by lady bird	9	0	0	1	0	0
Eats entire boll—no entire boll eaten	10	0	1	0	0	0

Source: Compiled by the Author.

These views were considered in selecting repertory-grid for environmental perception and behavioural studies in agricultural geography in developing countries.

Techniques for Analysing Repertory-grid Data

There is evidence that repertory-grid data are increasingly being analysed by factor analysis techniques (Slater, 1977), whilst Multi-Dimensional Scaling (MDS) has been virtually ignored as a possible alternative (Baker, 1977). Below is a summary of some of the implications of the assumptions underlying the application of factor analysis to individual repertory-grids (Baker and Richards, 1978):

1. A factor analysis of a grid utilises the analysis models, however, variables are assumed to be linearly related, yet there is no a priori reason why an individual's constructs should be linearly related. Indeed there is little evidence of any kind to suggest how they are related to help decide which mathematical transformations might be implemented.
2. Factor analysis assumes that its input variables are measured in ways that are statistically independent. Such an assumption is valid in many research contexts in psychology where a respondent is required to complete a series of separate "tests". Under Kelly's theoretical ideas, personal constructs are meshed together in a system and so it is difficult to imagine how an individual, in the course of grid completion, can consider one construct in isolation from his other constructs as he assigns scores to his elements.
3. Another assumption is that each variable is derived from a parent distribution which is normal. This implies that the extreme locations, that is, the poles on each construct are the least likely position to which a person will assign or rate his environmental elements. Whilst there is always a possibility of experimental bias-producing responses which tend to assume that each position along the construct is equiprobable for the assigning task to be valid. Indeed, if extreme pole positions are relatively infrequent for a set of elements which, presumably, lie within a construct's range of convenience, then doubt might be cast on the utility of the construct as a meaningful discriminator. Certainly acceptance of the validity of the normality assumption for grid data presumes constructs which are fundamentally different from Kelly's original notion of bi-polar discriminants.
4. Factor analysis of grids usually operate on data measured at the interval or ratio level. These data are derived from grid completion strategies which use rating scales. Numerical values or positions along the scale are supposed to indicate the

"amount" or location of a grid element along a discriminatory construct. There is a danger of a cultural bias in employing such strategies in an unfamiliar social environment; it is a moot point whether the act of assigning numbers or allocating scale positions has a similar meaning to a farmer in rural areas, say, in Africa or elsewhere in the world.

Non-metric MDS is suggested as an alternative to factor analysis where it is thought useful to analyse a grid by a relatively sophisticated technique which produces a multi-dimensional solution. But MDS utilises data which are measured at only a nominal or ordinal scale, thus simplifying the task of grid completion by reducing the time spent to elicit grids. Consequently, the application of rating scales is eliminated while the respondents are simply asked to allocate grid elements to either pole of a construct. To date, the use of non-parametric data effectively circumvents the objections raised on the appropriateness of factor analysis. MDS, like factor analysis, produces a simplified multi-dimensional model of the original data, but, typically, provides acceptable solutions in only two or three dimensions, fewer than is normal in factor analysis. But the higher dimensionality of factor analysis solutions can be ascribed to their being constrained by the linearity assumption. Nevertheless, the attraction of two or three dimensional graphical interpretations of grids should not be underestimated. This then raises the question of whether statistical analysis of grids both of the factor analysis and MDS type is essential or desirable. Certainly the issues need to be examined thoroughly otherwise an automatic and uncritical application of factor analysis becomes the norms in the repertory-grid technique, for example, both techniques assume that statistical relationships which are detected in a grid represent psychological relationships with an individual's construct system.

The factors obtained from individual grids are usually assumed to be in some unspecified way "superordinate". Similar interpretations could be given to the dimensions output by an MDS package. However, "superordinancy" of factors is hardly related back to hypotheses about the form of that particular individual's construct system, nor are re-tests made to see whether he recognises or uses superordinate constructs. This latter point is conspicuous by its absence in factor analysis which specifically involve hypotheses about the data variability and the nature of the factors, by partitioning variance into common and unique portions. Also, when elicited grids are small, for example, in studying farmers' perception of pests, there seems little sense in progressing beyond simple techniques (Frensella and Bennister, 1977) which can adequately illustrate the principal features of a grid.

Merits of Repertory-grid Techniques

1. Repertory-grid techniques have the qualitative advantages of survey work without the undesirable imposition of the investigator's opinions and perceptions.
2. The repertory-grid techniques are extremely flexible, allowing studies at the personal level as well as at the group level.
3. The results are often very stimulating in suggesting ideas, other research might pursue.
4. Repertory-grid techniques allow analysis of the relationship between a respondent's constructs by factor analysis or principal components analysis in order to identify the major principles on which an individual discriminates between phenomena. This confirms that there is potential for sophisticated analysis with repertory-grids.
5. The possibility of visualising a construct as a straight line from an idea to its contrary along which the individual will scale a given element is mathematically very convenient; various correlations and angular relationships can be studied. In addition, there is considerable experience in their use and analysis in psychology, and the mathematics are well understood.
6. The manner in which an individual relates certain constructs and elements to each other can tell us much, not only about his perceived world, but also about the personal meaning he attaches to the words he is using, the elicitation of these patterns of elements and constructs thus becomes an invaluable tool in communications, which is its prime use in psychiatry and offers much to behavioural geographers.

Limitations of Repertory-grid Techniques

1. The repertory-grid technique presents more difficulties to the interviewers than to the respondents. It is a relatively complex

procedure for interviewers to fill in the grid accurately as well as randomly present successive triads. Interviewers, therefore, need considerable training in the use and rationale of the grid before using it in the field.

2. The repertory-grid is an important means of probing categorisation, but it tends to provide only one possible taxonomy, whereas most people and cultures operate with a complex set of criteria for their environment.
3. The results are, generally, exploratory, although they may indicate possible areas of research to be pursued.
4. Although the constructs are those used by the respondent himself, it is the researcher who chooses which ones to use.
5. Grids are mainly seen as potentially an educational tool since they can encourage people to express their environmental experience and knowledge in novel ways, and thus possibly foster a greater concern for and participation in the discussion and solutions to their agricultural problems. The utility of repertory-grid methodology in developing countries is more closely in tune with Kelly's original purpose than geographical applications.
6. The technique is time-consuming, and probably most conveniently performed on a volunteer or captive population, for example, students. To complete a full grid probably requires one or more hours and this is a significant problem in using the repertory-grid as one part of a longer interview. Its complexity and time-consumption make this technique less useful in a survey or sampling situation.
7. Analysis is difficult, as it is hard to generalise from the individual, with his personalised constructs and elements, to the group as a whole.
8. There is often a danger of concentrating on grid analysis per se, and losing sight of the purpose and utility of such behavioural research. In dealing with farmers' problems in Africa, the use of repertory-grids is considered as only a small fraction of a more general practical and problem-oriented approach, rather than as an intellectual exercise in image measurement.
9. A mixed coding convention in repertory-grids may cause an analytical and processing problem depending on the association measure used. Thus, interviewer intervention in labelling may be inevitable but equally may introduce observer "bias".

10. Both factor analysis and MDS assume that statistical relationships which we detect in a grid represent psychological relationships with an individual's construct system. This assumption is always not true.
11. Also, when elicited grids are small, which is often the case, for example, in studying farmers' perception of pests, there seems little point in progressing beyond simple techniques which can adequately illustrate the principal features of a grid.

Summary and Concluding Observations

As part of his broader theoretical framework, Kelly proposed the repertory-grid technique, which has been used further in several investigations in social science. It is evident from the discussion that the technique is essentially an individual approach, though it can be adopted for group sessions. But its purpose is to provide a "map" of the respondent's personal constructs and their inter-relationships or changes over time, thus it is an attempt to find out how each person perceives his environment or at least that part of his environment with which the investigation is concerned, for example, farm problems. Repertory-grid testing is not a single test since it is a highly flexible technique which can be developed afresh for each new investigation.

In accordance with Kelly's theoretical approach, a construct is a way in which two things are alike different from a third one. The most important aspects of the repertory-grid technique are the constructs and elements (pests, farms, weeds, photographs, cards, etc.) to be included. But these depend entirely on the purpose of the investigation and should flow directly from it. Such purposes can, in turn, be classified as being concerned with elements, with structure, or with change. Thus, variations in the technique are often introduced, for example, the respondent may be asked to produce his own objects or the researcher may offer certain constructs to the respondent, rather than allow him to suggest them. This latter procedure has the merit of turning the repertory-grid into a group technique, so that many respondents can be asked to fill in the same grid, and factor analysis as well as other appropriate statistical techniques can be applied. When using the grid as a group technique, the analysis can be simplified by instructing respondents that they must give checks to half the number of elements on each construct. It the elements are photographs, cards or pests in bottles, the respondent simply has to sort these into two equal parts :

those that conform and those that do not conform, possess the construct in question.

The repertory-grid technique is extremely flexible as already noted, allowing studies at the personal level as well as at the group level. But it has not been applied widely in environmental perception and behavioural approach in agricultural geography. The present paper has attempted to deal with the problem of measuring attitudes and conceptualising farmers' perception of the environment. The discussion cited specific research field, the utility of traditional farmers' environmental knowledge in contributing to systems for monitoring ecological changes.

It is clear that both ethno-scientific studies and recent work in behavioural geography and psychology are concerned with the representation and evaluation of people's cognitive structures of the environment, and they share common methodological problems. Consequently, an approach to eliciting information from traditional farmers is envisaged which incorporates features of both since it is necessary to interpret knowledge within the context of their environmental images (Baker, 1977). In addition the use of PCT and repertory-grids is a valuable starting point because of their strong theoretical underpinnings, an integrated methodology and appropriate elicitation techniques. To date, the use of these methods in the context of the developing world could be strengthened by incorporating some of the techniques used in ethno-science, for example, MDS and careful lexicographic compilation of folk taxonomies for possible use as grid elements.

Apart from discussing PCT and repertory-grids, the paper has briefly examined the issues surrounding the application of MDS techniques and factor analysis in representing farmers' environmental perception. Above all, if a geographer may comment outside geography, the repertory-grid methodology can be used, as in this article, as a clinical psychologist uses it: not only to identify weaknesses and aberrations but to understand the positive, constructive aspects of the world view of a human group or individual, so that it is possible to deploy these rather than to medicate the symptoms of failure (Townsend, 1977). One of its greatest potentials appears to be in comparative studies where the perceived phenomena recorded by the repertory-grid can be measured against spatial data, for example, agricultural landuse and other functions of actual human behaviour. The technique will also be a significant instrument in the initial phases of research work, where elicitation of constructs alone can be of major help in research design, including work where survey research is appropriate in later phases. It is evident that the repertory-grid technique offers considerable utility in bringing out the multidimensional nature of the community environmental knowledge, for example, where there is a utility dimension to a plant or insect normally regarded as a weed or pest respectively or where evaluations vary between men, women, children or between different class and occupational groups, for example, farmers on the one hand and extension agents on the other. The technique also appears to have considerable utility in triggering off useful discussion and self-diagnosis, leading to heightened awareness of variations, in responses between individuals or groups and to proposals for minimising the conflict and misunderstanding engendered. Available evidence reveals that the approach discussed in this paper has a good deal to offer where there is genuine determination to "democratise" the development planning and rural education process and where "participatory" styles of environmental resource management are the ultimate aim. Nevertheless, the repertory-grid technique has a number of merits and limitations as summarised in this article.

REFERENCES

Baker, D., 1977 : *Some Methodological Issues in the Measurement, Analysis and Evaluation of Peasant Farmers Knowledge of Their Environment,* A Research Memorandum, MARC Report Number 9, Chelsea College, London.

Baker, D., et al, 1977 : *The Utility of the Nigerian Peasant Farmers' Knowledge in the Management and Monitoring of Agricultural Resources,* Monitoring and Assessment Research Centre, Chelsea College, University of London, General Report No. 4

Baker, D. and Richards, P., 1978 : *Repertory-Grid Methods and Environmental Images in Rural Africa,* IBG Quantitative Methods Study Group, SOAS.

Bannister, D. and Mair, J. M. M., 1968 : *The Evaluation of Personal Constructs,* Academic Press, London.

Bannister, D. and Frensella, F., 1974 : *Inquiring Man : The Theory of Personal Constructs,* Penguin Education.

Chetwynd, S. J., 1974 : *Generalised Grid Technique,* St. George's Hospital (Mimeo).

Dixon, C. J. and Leach, B., 1978 : *Questionnaires and Interviews in Geographical Research,* CATMOG 18, Geo Abstracts, University of East Anglia.

Frensella, F. and Bannister, D., 1977 : *A Manual for Repertory-Grid Techniques,* London.

Harrison, J. A. and Sarre, P., 1977 : Personal Construct Theory in the Measurement of Environmental Images. Problems and Methods. *Enviro. Behav.* 3, pp. 351-74.

———, 1975 : Personal Construct Theory in the Measurement of Environmental Images : Applications. *Enviro. Behav.* 7, pp. 3, 58.

Hudson, R., 1974 : *Images of Some Town in Country Durham : An Example of the Use of Repertory-Grid Methodology, North East Area Study Working Paper No. 12,* University of Durham.

Kelly, G. A., 1955 : *The Psychology of Personal Constructs,* Norton, New York.

Lundeen R., 1972 : *The Semantic Differential Technique and Personal Construct Theory in Image Measurement, Discussion Paper No. 5,* Department of Geography, York University, Toronto.

Oguntoyinobo, J. S. and Richards, P., 1978 : Drought and the Nigerian Farmer, *Journal of Arid Environment, 1.*

Oppenheim, A. N., 1966 : *Questionnaire Design and Attitude Measurement,* Heinmann, London.

Slater, P., 1974 : *Composite Diagrams and Systems of Angular Relationships Applying to Grids,* Institute of Psychiatry, London (Mimeo).

———, 1977 : *The Measurement of Inter-Personal Space by Grid Technique,* London (2v.)

Townsend, J. G., 1976 : Farm Failures : The Application of Personal Constructs in the Tropical Rainforest, *Area* 8, pp. 219-222.

———, 1977 : Perceived Worlds of the Colonists of Tropical Rainforest, Colombia, *Transactions,* Vol. 2, No. 4, Institute of British Geographers.

Whyte, A., 1977 : *Guidelines for Field Studies in Environmental Perception,* MAB Technical Note 5, UNESCO, Paris.

2

PERCEPTION OF DROUGHT HAZARD AMONG THE FARMERS OF INDIAN DESERT

A. K. TEWARI

DESPITE INJECTING huge sums of money in combating droughts in the desert and semi-desert areas of India, the paradox of human suffering has grown recently. We have to learn to cope with such devastating events from the actual experiences of the farmers who can mitigate the impact of calamity in their own way. Planners' inability to incorporate the folk perception has brought frequent miseries and successive famines to the arid-land farmers whose economy is totally shattered during and after such events. This is more evident in the desert fringe villages where despite great physical susceptibility to rain failure, the rural density of population is high. Perception is a four-stage psychological process of human behaviour where an individual senses a situation in his own perspective, thinks of it, remembers it and finally reacts to it. Accuracy of perception makes an individual better prepared for and adjusted to a hazardous situation. How far this behavioural process can be made useful in facing and adjusting to environmental risks is the major question dealt with in this paper.

The study area lies in the south-eastern margin of the Great Indian Desert sprawling over the districts of Jodhpur, Pali, Barmer, Jalore and Sirohi of Rajasthan. It is commonly known as the Luni Basin after the river Luni which drains this small sub-region (Fig. 2.1). During July and August of 1979 this part of the Indian desert was an arena of flash floods while it is still in a grip of severe drought. The paper aims at analysing the inhabitants' impression obtained by a questionnaire survey of 751 households in five selected villages representing the three geoeconomic gradients of a desert region, namely, arid, semi-arid and sub-humid. Spatial variations in hazard perception have been analysed in terms of awareness of drought, severity, frequency of occurrence, variations in the degree of personal experience and type of farm operation. The central theme is to evaluate the farmers' conceptualisation of the problem and their instant connotations to uncertain events.

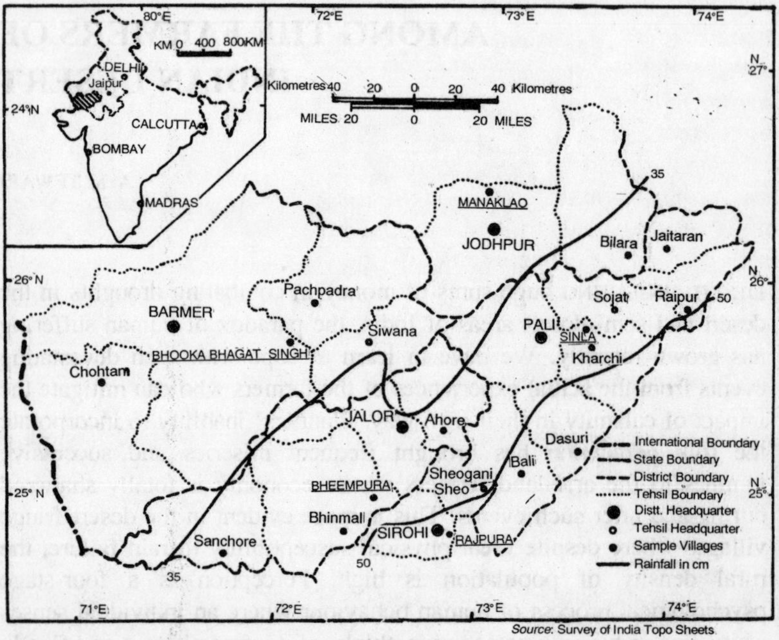

FIG. 2.1: Luni Basin — Location of sample villages.

Drought Situation in the Study Area

The area of study is characterised by hyperthermal conditions and low and erratic rainfall. It remains under the severe grip of drought after every three to four years. Its resource potentials are limited except in the irrigated areas. The high rural density of population especially in the eastern parts stands as a contrast to marginal yields from farm and pasture which fluctuate with the cycle of rainfall. The largest record of rainfall has failed to reveal any cyclic recurrence of drought conditions. In the worst years of scarcity in the last century, thousands of people have died from starvation in the years 1812, 1837, 1868 and 1899. The impact of drought in the present century is confined to mass exodus of the lower section of rural society and general malnutrition of the entire population. The severe droughts of 1966-67 and 1973-74 prompted the Government to set up permanent planning agencies for drought relief and rural works. These were later redesignated as Drought Prone Area Programme (DPAP). In the present study five severely affected villages have been selected in such a manner that all the three strata of climatic gradients—arid, semi-arid and sub-humid—could be represented. Scarcity of water, grass and grain has become so much a part of the desert farmers that it excites no comment until these three calamities are felt together. According to a local proverb drought and famine are synonymous with this region (Adams, 1899; Tod, 1920). The geographic characteristics of the sample villages are given in Tables 2.1, 2.2, 2.3 and 2.4.

TABLE 2.1: **Categories of sample villages**

Sl. No.	Situation	CG	EG	Population* 1961	1971	1977	No. of HH surveyed	% Population surveyed
1.	Lat. 26°27′N Long. 73°0′E	Arid	I	897	1497	1179	192	78.75
2.	BS Lat. 25°40′N Long. 72°0′E	Arid	U	1461	1663	1230	216	74.86
3.	Lat. 25°47′N Long. 73°35′E	Semi	I	1244	1563	1137	216	72.74
4.	Lat. 25°7′N Long. 72°24′E	Semi	U	519	647	365	55	56.40
5.	Lat. 24°53′N Long. 72°54′E	Sub	I	318	464	372	72	80.17

* Population interviewed; CG=Climatic Gradient; EG=Economic Gradient; Semi=Semi-arid; Sub=Sub-humid; I=Irrigated Villages; U=Unirrigated Villages; HH=Households.

Drought and the Strategy of Perception

Droughts have been defined as the number of days receiving less than the specific amount of rain or certain percentage of normal rainfall for the period concerned (Subramanyan, 1958). It differs from other types of natural hazards in terms of its variable duration and indefinite beginning or end. Its perception also varies from person to person depending on the geographical location, past experience and its effect on major economic pursuits. It also varies with farmers operating on irrigated fields and unirrigated fields, farmers with cereals or single-crop emphasis and farmers with cash crop or multiple crop emphasis. Thus, the conception of drought of a farmer depends on the timeliness and nature of shower, i.e., rainfall effectivity. Thus a more meaningful conception of drought would be moisture stress days for a particular crop as determined by soil moisture supply, potential evapotranspiration and crop root zone. An analysis of drought indices on the regional or sub-regional scale must incorporate variable biological responses like the nature of crop, its stage and rooting depth to the meteorological parameters. Balmer (1965) has developed a climatic analysis system permitting time and space comparison to evaluate a meteorological anomaly characterised by a prolonged and abnormal moisture deficiency. In the sampled villages the mean growing period with normal rainfall of more than 5 mm during the monsoon season (June to October) is hardly 11 weeks. If a farmer fails to keep to the timing of sowing his crops to this limited period, he comes under the grip of drought. Here the role of perception as a guiding force for day-to-day agricultural activity becomes self-evident. The farmers of this region are aware of the probability of drought over a long period but when exactly the lean year will occur is uncertain. Moreover, the increase in pumping of ground water for cash crops creates man-made drought for the smaller farmers who cannot afford tube-well irrigation.

TABLE 2.2: Physical attributes of sample villages

Sl. No.	Distance (km.) from nearest					Elevation in metres	Relief	Water bodies	Depth of water table
	City		Railway station		Metalled road				
1.	Jodhpur	22	Manaklao	2	2	246	Sandy Plain	Seasonal Stream	40
2.	Balotra	24	Balotra	24	1	100	Sand Dunes	Salt Basin	10
3.	Pali	30	Marwar Jn.	8	2	180	Sandy Plain	Seasonal Stream	15
4.	Bhinmal	19	Sheempura	0	1	170	Sandy Plain	Rain Water Tank	20
5.	Sirohi	8	Sirohi Rd.	15	1	330	A Valley surrounded by Aravalli	Seasonal Stream	25

TABLE 2.3: Bio-climatic profile

Sl. No.	Vegetation	Soil	Recording Station	Climate				
				M	T	H	R	RD
1.	Scrub thorn xerophytes and acacia (*senegal*)	Red brown desert soil with sandy loam structure	Jodhpur	–45	24	53	34	20
2.	Scrub thorn xerophytes and acacia (*senegal*)	Yellow brown desert soil	Pachpadra	–49	26	56	27	13
3.	Thorn trees with desert grasses (*dichanthium elyourus*)	Red brown desert soil	Pali	–43	23	55	41	19
4.	Thorn trees with desert grasses (*dichanthium elyourus*)	Red brown desert soil	Bhinmal	–40	24	52	47	17
5.	Thorn trees with grass cover (*themedapennisetum*)	Yellow Sandy foothill loam	Sirohi	–29	21	54	58	26

Note: M = Thornthwaite's Moisture Index; T = Temperature °C; H = Humidity %; R = Rainfall c.m.; RD = Rainy Days in one year (all average of 40 years).

TABLE 2.4: Landuse in sample villages — 1971

(Area in hectares)

Sl. No.	Forest	Cultivated Land						Fallows	Waste land	Area not available for cultivation	Total area of the Village
		Irrigated			Unirrigated						
		K	R	S	K	R	S				
1.	0	129	34	well	1682	—	—	846	65	111	727
2.	0	—	—	—	3684	—	—	1760	1936	168	7563
3.	0	227	6	well tank	1432	1	—	1671	115	136	3621
4.	0	—	—	—	367	—	—	108	200	29	703
5.	8	16	4	well	237	7	—	153	147	382	969

Note: K = Kharif season (monsoon period); R = Rabi season (Winter); S=Source of irrigation; Fallows include both old and new fallow lands; Wasteland includes cultivable waste, grazing lands, rough pastures and fallows not cultivated for more than five years; Area not available for cultivation includes all village land, occupied houses, settlements, harvesting ground, roads, railway lines, burial grounds, tanks, wells, etc.

Perception and Hazard Management

Man and environment interaction is a fundamental concern in all academic fields dealing with the problems of human welfare. This perennial topic is so open-ended that only through interdisciplinary perspective, elucidation and illustration, we can arrive at a solution. Man may be an ecological dominant in favourable areas but in areas of adverse environment his adaptive role, his capability to structure and order environment intellectually according to his design of social manifestation are more important (Tuan, 1977). In the arid and semi-arid areas of India the farmers operating on unirrigated fields submit to the all-pervasive and overwhelming forces of climate and only through adaptive responses and ritual adjustments they survive and grow. The cycle of drought comes to them as a dramatic event when overgrazing of pastures and overtilling of fields may be disastrous not only for the core areas but also the lands bordering the deserts. The soil depletion and desertification always follow severe droughts. Man's sense of affinity with the natural order is very strong in areas of uncertain environment (Burton, 1978). The emphatic and intellectual understanding of the hazard makes the inhabitants feel at ease. They can adjust to a problem wisely according to their own sense of impression.

The concept of perception has many useful applications in the field of natural resource management and folk decision-making (Kates, 1976). It is considered as a necessary prelude to improving hazardous situation. It is hypothesised that a farmer's understanding of his immediate surrounding plays a significant role in his individual and collective decision to adopt any practice that can make his environment productive, reliable and safe from eventual risk (Porteous, 1977; Proshansky, 1976; Saarinen, 1966). Here in the study area the only problem is due to partial or complete failure of monsoon rains after every three to four years. It leaves millions of natives struggling for their bare existence. The Government during this period provides temporary subsidies of grain for work. The works so undertaken in rural areas, such as road construction and building of government houses are mostly unproductive. The Drought Prone Area Programme of the Government, has, no doubt, taken over the task of structural development of such areas, but it has not endeavoured to ascertain the views and ideas of the inhabitants who face this problem. We have a reasonably good amount of information about rural economy of such areas but we know very little about the reactions of farmers to the stress and strain of their major problems. Clearly the arid land resources cannot be adequately managed until the development programmes are in conformity with the aspirations and goals of the local residents. The major thrust of developmental endeavours in such areas may have three distinct frames: ecological frame, i.e. enhancing the water supply through techno-scientific means (inter-basin transfer of surface water such as the Rajasthan Canal Project); economic frame, i.e. to increase the profitability of existing resources (substitution of one activity by another); and behavioural frame, i.e. inhabitants' understanding and using their environment as an aid to adjustment to any eventual hazard. If the farm perception of the farmers can guide the planning of landuse, choice of crops and inputs, farm insurance, etc., it can be of immense relevance for the indigenous rural management (White, 1974).

The agricultural organisation in the sampled villages is predominantly owner-operated family enterprise with farm holdings ranging in size from 0.8 hectares to 8 hectares (Table 2.5). Such farmers can neither take any risk nor adopt new production techniques until they conform to their own or family knowledge of farming. Thus in a peasant farm family with a fixed amount of land (mostly unirrigated) and a fixed number of family labour, the postulation of economic behaviour is confined to maximising the immediate food requirement of their family by choosing traditional crops of their preference and proven success. This effort maintains the status quo of rural landscape. Bound by constraints and customs the arid land farmers are excessively cautious in adopting innovative techniques recommended by non-farmers. Their perceived opportunity to innovate lies only in their individual or group experience of the recent past, learning through the painful events of drought and famines. If a poor farmer of a Hindu community is asked to diversify his rural economy by adopting poultry of hog-raising, his sentiments are hurt. He will prefer to calculate the rule of thumb basis the past experience of his problem and will adopt landuse of his own guidance. The subsistence farmers prefer to grow such crops which give the highest minimum yield under any weather condition. In the Luni Basin the farmers cultivate pearl millets (Bajra — *Pennisetum typhoideumpers*) every year irrespective of good or bad monsoon, or the price slump in the open market. Their perception guides the day and time of sowing, ploughing and harvesting. They are, no doubt, less susceptible to

TABLE 2.5: Characteristics of farm units

Sl. No.	Households						Cultivated Land Type			Holding in hectares	MGS (Kg.)	Livestock per person		Resi- dency years	Income per capita (Rs.)	Ave- rage age (yrs.)	Facili- ties
	F	C/DR	O/DR	L	W	MEQ	A	B	C			Cattle	(Sh)				
1.	5.9	76	5	15	463	840	9	8	83	2	100	1	3	38	300	40	4
2.	6.2	94	4	5	361	800	—	1	99	8	145	1.2	5	40	180	43	3
3.	6.0	78	7	14	484	865	6	8	86	4	129	1.1	4	41	310	42	4
4.	6.1	80	8	16	239	135	—	1	99	6	130	1.2	4	39	280	41	4
5.	5.8	68	5	6	125	252	1	6	93	1	80	1	2	37	235	38	2

Notes: F = Number of persons in one family. C/DR = Child dependency ratio; O/DR = Old dependency ratio; L = Per cent population literate; W = Number of workers; MEQ = Man equivalent labour force; one adult male=one unit; one adult female = 0.75 unit; and one child or old 0.5 unit; Cultivated land type; A = per cent area irrigated, B = per cent area semi-irrigated, C = per cent area unirrigated; MGS = Median grain surplus in good year of rainfall, kilogram per person; Facilities = weightage of necessary facilities available in one household, average of sum of values as per: Cow = 3, Bullock = 3, buffalo = 4, Sheep or Goat = 1, Camel = 8, Bullockcart = 5, Plough = 5; Well = 5; Tractor = 20.

drought conditions and are able to risk aversion, yet remain poor. In the present paper simple hypothesis of man: drought relationship has been formulated and set up for testing on the basis of empirical data collected from the field. It will be seen that the process of perception varies according to the aridity of area and amount and experience of severe droughts.

Perception of Drought

The kind of problem faced by the farmers of the arid lands of India is unique in itself because the population has increased in recent years without any perceptible shift in the occupational structure. Agriculture-cum animal husbandry are the mainstay of their economy. The unorganised proliferation of agriculture in a water-deficit zone is also a major cause of desertification. Most of the farmers perceive drought (dry spell) as the only disadvantage of their villages. Perception of the present moisture condition is also somewhat accurate. Since no data of rainfall and soil moisture is available for the locales of inquiry, the author has based his findings on the meteorological data available for the nearest recording station. Palmer's drought index has been applied for comparative purposes. It is based on monthly values of moisture deficiency/excess and the amount of moisture required for normal weather in each month (Palmer, 1965). Thus it is the departure of moisture from the normal condition as expressed in terms of numerical indices (Fieldhouse, 1965). Positive values indicate wetter than normal conditions and negative values indicate drier conditions. This index of monthly values of drought severity in ascending order is expressed as near normal : 0.49 to -0.49, Incipient : -0.50 to -0.99, Mild : -1.1 to -1.99; Moderate : -2.00 to -2.99, Severe : -3.00 to -3.99, and extreme : -4.00 and above. The end of a drought period is determined by the equation :

$$P_e = \frac{100\,V}{Q}$$

where P_e is the probability that the drought has ended and is expressed as moisture received at the end of a drought. It is the potential evapotranspiration value expressed in percentage of the amount required. V indicates the time when the amount of wetness effective in ending a drought becomes positive. This method is applied for the determination of drought severity in the United States and can well be applied for the drought-prone areas of India — if the index values are summarised for the growing period.

In perceiving the present moisture condition most of the farmers replied according to the rainfall useful to their crops. In the extremely arid village (Manaklal and Bhooka Bhagat Singh) their assessment was very close to the actual conditions, whereas in the semi-arid and sub-humid village it was variable (Table 2.6). And indeed the worst droughts are more misery-causing in the desert margins rather than in the hard-core desert itself. Normally a drought cannot attain extreme severity in less than four months. If farmers can understand the incipient drought period correctly they can save their crops from complete failure by adopting certain measures in advance. The farmers who had sown their field just after the first shower (June, 1977) were not as satisfied with their crop prospects as those who preferred to wait for the next shower. The latter's argument was that if the rainfall was above normal for the last two successive years (as the case was) there will be an erratic behaviour in the following years. And really it so happened that there were long dry spells between the first heavy shower and the next showers in several villages.

TABLE 2.6: Perception of drought

Villages	Percentage of farmers who perceived as main disadvantage of their fields			Farmers' assessment of present moisture condition (% No.)			
	Aridity	Soil infertility	Others (winds)	Drier	Average	Wetter	Actual
1. Manaklal	90	6	4	30	60	10	N
2. Bhooka Bhagat Singh	80	3	17	12	81	7	N
3. Sinla	95	3	2	40	81	7	N
4. Bheempura	80	10	10	58	31	11	I
5. Rajpura	70	15	15	65	30	5	I

Note: Actual drought condition in Palmer's Index for the nearest recording station. N = Normal; I = Insipient.

TABLE 2.7: **Comparison of farmers estimation of severe drought frequency with actual drought condition for the last 35 years**

	Villages	% Years when farmers consider good crops	Actual % of years of good crops*	% years of drought perceived	Actual % of years of drought**	Average yield of millets+
1.	Manaklal	20	35	25	16	4
2.	Bhooka Bhagat Singh	25	30	35	25	3
3.	Sinla	23	38	34	21	6
4.	Bheempura	31	42	36	24	6
5.	Rajpura	52	62	34	28	7

* Based on revenue records.
** Based on records of food and fodder shortage.
\+ Average yield of the village in 1976-77 in quintals per acre.

Actual Drought and Perceived Drought

The actual frequency of drought years for the last 35 years was slightly less than the farmers perceived. When the farmers were asked to estimate the number of years they faced dry spells and their harvest was poor, their normal reply was "formerly there were more frequent droughts than today". About 75 per cent of the middle-aged farmers mentioned 1941, 1949, 1960, 1966 and 1972 as bad years. The actual percentage of drought years in Table 2.7 has been computed from the records of the village *patwari* (a Government servant who keeps village records) for the last 35 years. This corresponds with the period when most of the farmers had gained their farming experience. The overestimation of past miseries is due to the general belief of the elderly people that they lived in harder days than their younger counterparts. This tendency to overestimate both good and bad harvest is inherent in human behaviour.

Role of Drought Experience in Perception of Hazards

Perception of any hazard improves with increasing experience unless a respondent fails to recall past events or becomes sceptical due to the problems of old age. In the village surveyed 41 per cent of the population belonged to the lower age group, only 3 per cent to the old age group and the rest of the population to the middle age group. In developing a hypothesis of relationship between drought experience and perception, it is necessary to take into account the total number of months of moisture deficiency during the crucial period of crop growth. Moreover, the response of very old farmers should also be excluded in arriving at any conclusion (Saarinen, 1966). Despite greater experience very old people have their own idiosyncracies and consider the vagaries of nature as a usual phenomena not worth noticing. In Tables 2.8 and 2.9, perception of drought has been calculated from the farmers' estimation of the number of drought years during a course of 35 years as compared to the actual drought frequencies derived from the records of crop yields of the respective villages. The more experienced farmers have a better assessment of drought risk, but it does not materially help them when faced with uncertainty, i.e. the estimation of drought in a particular year. Thus most of the farmers fail to make any grain and fodder reserve for the lean period. The only reserve farmers are habitual of making during a good harvest is in the form of gold, silver and brass ornaments or purchasing greater number of livestock and cattle. Marriages and other social ceremonies are also performed during the years of good crop yield. During the years of drought the barter system within a village starts with the exploitation of the lowest strata of rural society. The smaller and marginal farmers start selling their stocked commodities in a certain order; first their cattle, then their ornaments and utensils and when everything is exchanged they migrate to other localities till the next monsoon. In the absence of any banking facility in the village, the practice of purchasing gold and silver from the village *baniya* (trader) during good harvest and exchanging the same for food and fodder at the time of drought is continuing. If the perception of drought is good the farmers can be saved from this natural exploitation.

TABLE 2.8: **Drought experience and perception of drought risk**

Villages	Number of months of actual drought experience					
	<60		60-120		>120	
	A	P	A	P	A	P
1. Manaklal	40	L	56	M	55	H
2. Bhooka Bhagat Singh	37	M	43	M	51	L
3. Sinla	38	L	50	H	54	H
4. Bheempura	42	M	45	M	50	H
5. Rajpura	39	L	44	M	56	M

Note: A = Average age; P = Perception; L = Low; M = Medium; H = High; V = Variable.

TABLE 2.9: **Measures of drought experience and perceived loss or stress**

Measures Perceived	Village Nos.				
	1	2	3	4	5
Recurrence: Median farmer's estimation of bad years in % of years	40	55	38	42	30
Loss: % farmers reporting substantial damage to their crops and cattle	60	76	65	62	54
Stress: % farmers reporting common family experience of hunger	15	20	32	23	28

When the farmers were asked to comment if there were any chances of drought in the next year or they had any idea of a bad harvest beforehand, their replies showed their farming experience had some relationship with perception. The elderly group had an overestimation of drought risk while the younger group had underestimated the future strains. Some of the new dry farming techniques and uses of chemical fertilisers have also influenced the younger generation but their percentage is very low. The spatial variation in the estimation of the recurrence of drought and its effect on economy has some correlation with the aridity of the area (Table 2.9). In the extremely arid villages the estimation of recurrence and loss due to drought is higher whereas the estimation of family experience of hunger is lower. This is because of the fact that the dependence on agriculture is lower in the extremely arid localities and the farmers are better adapted to hazardous events.

Type of Farm Operation and Perception of Drought Hazard

Perception of drought varies with the type of farm operation. Out of 751 cultivators interviewed more than 75 per cent operated on straight grain mainly millets in the monsoon season (Kharif) and wheat in the winter season (Rabi). Those who cultivated only grains were more perceptive than those who cultivated different crops with livestock emphasis. This is because the former type of cultivators have to depend more on the amount and timeliness of rains than the latter. The livestock raisers and cattle breeders can move their herds when the fodder is not available but the grain farmers lose their reserve capital, family labour, and the price of seeds and fertilisers. Many of the professional animal grazers of western Rajasthan operate a thriving business of purchasing cattle and livestock at very cheap rates during the worst period of drought and selling it to other parts of India. Even these cattles are driven out of India due to open desert borders. To check this practice and save the interests of farmers the Government under the Drought Prone Area Programme (DPAP) gives interest-free loans to the marginal farmers; but the operative part of the programme is so uncertain that its fruits hardly reach the actual farmers at appropriate times. Thus it is clear that the livestock raisers are less perceptive yet remain unaffected from any direct loss or stress during droughts (Table 2.10). A few cultivators who operate on cash crops such as spices (chillies), etc., are also very perceptive to weather conditions because the rain failure has a direct impact on the water table and cost of pumping. The high initial cost of such an operation and the State taxes make its cultivation very risky in the arid areas. During the severe droughts the water table goes down and the tube-well owners are prohibited from using pumped water for irrigation. Thus it may be concluded that drought is the concern of those farmers who depend on weather directly and are also more perceptive of the hazard.

TABLE 2.10: **Percentage of drought hazard and type of farm operation**

Name of Village	Farm Operation					
	Most Perceptive			Less Perceptive		
	S	D	L	S	D	L
1. Manaklal	64	40	30	36	60	70
2. Bhooka Bhagat Singh	70	60	25	30	40	75
3. Sinla	65	48	28	35	52	72
4. Bheempura	68	57	30	32	33	70
5. Rajpura	72	48	29	28	52	71

Note: S = Straight grain mainly *bajra*; D = Diversified : grain with irrigated vegetables and spices; L = Livestock emphasis : cow, camel and goat.

Conclusion

Perception of and reaction to the vagaries of nature are necessary for the bare survival of the closed rural economy. The farmers of the sampled villages are aware of the hazard. They are also able to assess the present moisture condition accurately. In the village situated on the desert margin the hazardous events are not so frequent yet the impact of drought is high due to greater population density. This is the zone of real suffering. Longterm residents, more experienced farmers, and those whose economy is most vulnerable to the variations of the weather perceive the risk more realistically and are better adapted. In order to alleviate the problems of drought with a new perspective, there is a need of better appreciation of farmers' conceptualisation of the drought and their native measures to dilute its impact. Only physical development of these areas is not the real answer. The role of aridity in the perception of drought is also reflected in this area. The farmers of drier zones are more perceptible to the risk despite the fact that their economy is not wholly dependent on foodgrain production. Animal-rearing has been adapted by them after years of trial and error. They are happier in the long run than the farmers of the desert margin who fail to perceive that they are living in a zone of permanent hazard. This aspect be highlighted in the future developmental plan of the arid regions of India. Non-crop economies are the only permanent answer to the risks and strains of such areas.

REFERENCES

Adams, A., 1899 : *The Western Rajputana States: A Medico-topographical and General Account of Marwar, Sirohi and Jaisalmer,* Junior Army and Navy Stores, London.

Burton, Ian, Robert W. Kates and Gillbert F. White, 1978 : *The Environment as Hazard,* Oxford University Press, New York.

Fieldhouse, D. J. and W. C. Palmer, 1965 : "Meteorological Drought and Agricultural Drought", Bul. No. 353 of the University of Delaware, Nowrh Del.

Government of Rajasthan : *District Census Handbooks* 1972 : *Jodhpur, Barmer, Pali, Jalor and Sirohi,* Census of India, 1971, Series 18, Rajasthan Pt. 10A and 10B, Jaipur.

Kates, Robert W., 1976 : "Experiencing the Environment as Hazard", in *Experiencing the Environment* (Eds.), Seymor Wapner, Saul B. Cohen and Bernard Kaplan, New York, Plenum Press.

Palmer, W. C., 1965 : "Meteorological Drought", *Research Paper No. 45,* U.S. Weather Bureau, Washington D.C., Feb.

Porteous, J. Douglas, 1977 : *Environment and Behaviour,* Reading, Mass, Addition-Wesley Publishing Company.

Proshansky, Harold M., 1976 : William Ittleson and Leanne G. Rivlin, *Environmental Psychology,* New York, Holt Rinehart and Winston, Esc. Edition.

Saarinen, Thomas, F., 1966 : *Perception of Drought Hazard on the Great Plains,* Deptt. of Geography Research Paper No. 106, University of Chicago, Chicago.

Subramanyan, V. P., 1958 : "Drought and Aridity in India : A Climatic Study", *Proc. of I.M.D. Symp. on Met.*

Tod, James, 1920 : *Annals and Antiquities of Rajasthan,* Vol. I, 11, (ed.) William Crooke, Oxford University Press, London.

White, Gilbert, F. (Editor), 1974: *Natural Hazard: Local, National and Global,* Oxford Univ. Press, New York.

Tuan Yi Fu, 1977 : *Space and Place,* University of Minnesota Press, Minneapolis.

3

PERCEPTION OF DROUGHT AND DESERTIFICATION BY INDIAN DESERT DWELLERS

L. P. BHARARA

PERCEPTION OF drought and desertification is an important component of environmental perception in arid areas. It is used to mean human awareness and understanding of the causes and processes of this environmental phenomenon. It reveals man-environment relations in which man's individual and collective understanding of the environment is seen as a major focus in shaping that environment through the action of man's choices and behaviour. A perception approach to man-environment relations recognises that for each objective there are many perceived elements and relationships as seen and understood by different people and at different times and places. Man reaches decisions and takes action within the framework of his perceived sets of elements and relationships. The concept of perception includes much more than individual sensory perception. It reveals individual and group judgement, an evaluation of alternatives and the empirical study of decision-making and of policy options.

Studies on perception of environment for human behaviour, appeared in the 1950s or even earlier. But until the 1960s there was little recognition of it. As a multi-disciplinary field with some common research problems, environmental perception began to appear in the early 1960s. Earlier ideas about cultural influences on man's use of the environment became re-expressed in terms of the influence of the perceived or subjective environment. These ideas soon led to a rapid development of empirical investigations into environmental perception and the need for more knowledge about the role of environmental perception, in environmental management became identified as a practical objective. During the last decade, environmental perception has become a rapidly expanding areas of research that is empirically and practically oriented.

Work on environmental perception and specifically of natural hazards by psychologists and geographers or anthropological studies of folk taxonomies of environmental phenomenon, has recently developed. Various studies (Brookfield, 1969; Craik, 1970; Goodey, 1971; Lowenthal, 1967; Proshansky, et al., 1970; and Saarinen, 1966; '76) have dealt with the psychological approach to environmental behaviour. Perception literature with a more environmental resource management focus includes Burton and Kates (1964); O'Riordan (1971); Swell and Burton (1972); and Whyte (1977). These reveal perception and management of environmental hazards and guidelines for field studies in environmental perception. Recent preliminary studies (Bharara, 1978, 1980, 1982, 1986) in arid zone of Rajasthan relating to drought perception, recollection and prediction are lacking insight of drought and desertification problems perception of which need to be studied.

This paper seeks to make a contribution to the ethno-science of droughts. It investigates the memory of drought over a period extending back to the limits of living memory in the villages of western Rajasthan. An attempt is made to add to our understanding of how memory works in a cultural system which includes the synoptic perception on past droughts as all-encompassing phenomena — a perception of which is based the expectation and prediction of droughts to come. The material takes an additional significance from the changing context of drought, in consequence of population growth, technological development and economic and social change.

It is easy to forget that the categorisation of phenomena as for example, climate, biological and social, though convenient and useful, is an arbitrary product of a particular intellectual tradition. In some social situations, such as Rajasthan village communities in western Rajasthan, where these compartmentalisations of knowledge are not to be automatically made, drought is not simply the dearth of rain (though everyone knows that rain would remove it) — it is the total quality of life, including besides weather, animal behaviour and social relations. The value of this material will serve a key to understanding the coping mechanisms and potential reactions of a rural population to the droughts which are often harbingers of the severer and longer term process of desertification.

Methodology

The study was undertaken in the Arid Zone of Rajasthan. The arid zone of Rajasthan — the universe of study, comprises of 12 north-western districts of the state (Krishnan, 1968; CAZRI, 1979). The districts are namely Barmer, Bikaner, Churu, Ganganagar, Jaisalmer, Jalore, Jhunjhunu, Jodhpur, Nagaur, Pali, Sikar and Sirohi. Based on agro-meteorological situation and extent of rainfall, chiefly 11 districts excluding Sirohi were taken into account for the study purposes (DST, 1981; CAZRI, 1979). The period of study of intensive empirical field investigations relates to the years of 1984-86.

A multi-stage stratified sampling design was followed. First, the universe was stratified into three tracts based on the amount of average rainfall and the mean aridity index. Rainfall influencing aridity is the chief variable inducing drought and desertification in the area. The tracts demarcated were 'A' receiving an average annual rainfall of less than 200 mm having the mean aridity index 84 and consisting of the districts of Jaisalmer, Barmer, Bikaner and Ganganagar. The tract 'B' receiving an average rainfall of 200 mm to 400 mm and having the mean aridity index 75 constituted the districts of Jodhpur, Nagaur, Churu and Jalore. The districts of Jhunjhunu, Pali, Sikar and Sirohi constituted tract 'C' receiving the average annual rainfall of 400 mm or more and having the mean aridity index 58 (Malhotra and Bharara, 1968).

The tracts reveal differential ecological pattern. The predominant land form in the tract 'A' consists of sand dunes of 50 to 100 metres height which cover about three-fourths of the total area. The area under inter-dunal plains is, therefore, much less. However, there are some rocky out-crops and these rocky areas are mostly barren and devoid of soil cover except in low lying areas where some sediments collect. The water table in this tract is very deep–ranging from 30 to 60 metres and the water found is often brackish to saline. The predominant land forms in tract 'B' are alluvial and sandy plains with scattered sand dunes and rocky exposures. The interdunal plains are comparatively extensive. The depth of water table ranges from 15 to 30 metres. The texture of the soil even in the alluvial plains is, however, sandy. The tract 'C' has predominant land forms of sandunes, hilly out-crops and extensive alluvial plains. The density of sand dunes is comparatively less and sand dunes are also of smaller heights. The drainage net-work is comparatively greater. The depth of water table is comparatively shallow and does not exceed 30 metres. The quality of water is comparatively better. Socio-economic characteristics of the population inhabiting these tracts differ (Malhotra and Bharara, 1968).

Secondly from each tract one district was so chosen as to represent a different ecological pattern — areas of predominantly pastoral nomadism, predominantly rainfed and rainfed with irrigation facilities. These representative areas comprised of three districts i.e. Jaisalmer representing pastoral; Nagaur representing rainfed and Pali representing rainfed with irrigation facility.

Thirdly, from each selected district, 4 villages for intensive empirical field investigations were selected. In a pastoral tract the villages selected were Bandhewa, Bharewala, Khetolai and Sankda from Jaisalmer district. The average annual rainfall here is 165 mm which is quite less and, therefore, pastoral nomadism and migration is predominant. Village Bhojas, Chhajoli, Rohina and Rohini in district Nagaur were selected from the dryfarming tracts having high textured soil. The average annual rainfall here is 388 mm with rainfall cultivation as the main occupation. Villages Dhabar, Dhurasni, Jheentra and Sardar-Samand were selected in Pali district representing irrigated areas with winter/rabi crops. The average annual rainfall here is 490 mm suiting to minor irrigation facility in the area (Fig. 3.1).

The selection of these specific villages was done purposively on the basis of various characteristics representing the universe. The factors taken into consideration were :

(i) Religion —Hinduism and Islam;
(ii) Caste — Balloach, Panwar, Parihar (Sindhi Muslims): Bishnoi (Sect); Rajput, Jat and others; Rajput, Gujjar and Ghanchi, etc.
(iii) Occupational predominance — livestock rearing, agriculture. agro-livestock and others including crafts;
(iv) Size of village — small (< 150 households), medium (150-250), Large (250-300) and very large (300 and above); and
(v) Land Form Unit i.e. village's soil type — the chief source of sustenance of the local population which in turn depends upon it for its source of livelihood. The land forms of the villages selected were chiefly Flat Alluvial Plains, Dune-cum-interdunal plains, Dune-cum-alluvial plains and Rocky-cum-buried pediment.

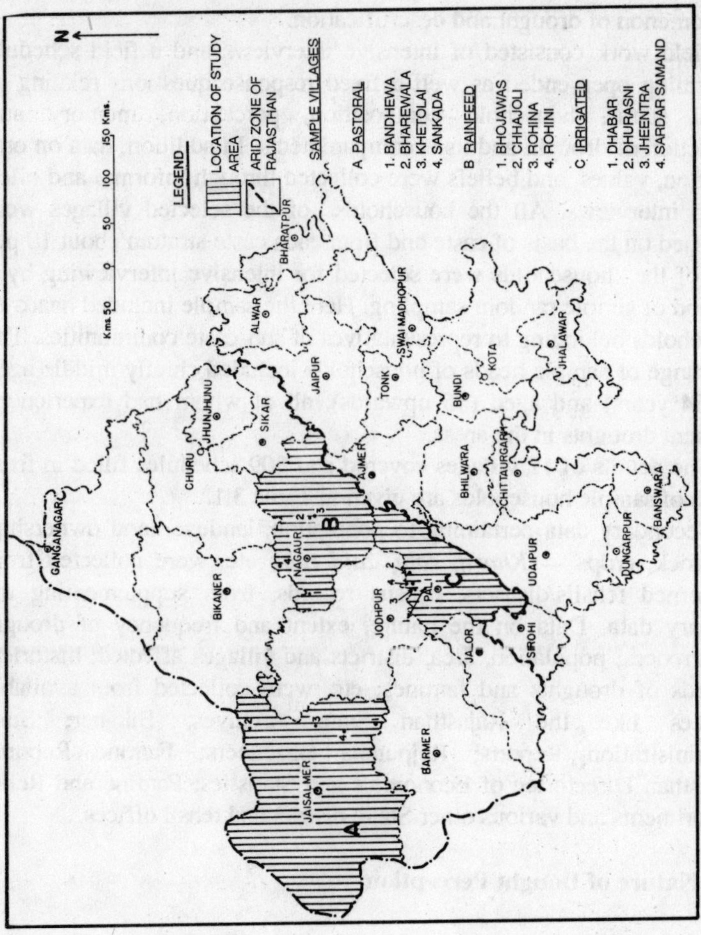

FIG. 3.1: Location of study area.

All these factors representing different ecological regions of the universe influence or are being influenced directly or indirectly by the phenomenon of drought and desertification.

Field work consisted of intensive interviews and a field schedule containing open-ended as well a fixed response questions relating to social status, household composition, perception, memory and prediction of drought and its accompaniments. In addition, data on oral tradition, values, and beliefs were collected through informal and often group interviews. All the households of the selected villages were stratified on the basis of caste and from each caste stratum about 10 per cent of the households were selected for intensive interviewing by a method of simple random sampling. Here the sample included heads of households belonging to representatives of sub-caste communities. The age range of sample heads of households included chiefly middle aged (35-54 years) and aged (55 upwards), all of whom had experienced frequent droughts in the areas.

The details of 12 villages covered and 309 schedules filled in from heads of sample households are given in Table 3.1.

Secondary data pertaining to population, landuse, land ownership, livestock, crops — *Kharif, rabi, zaid rabi,* etc. were collected from concerned tehsils/districts, census records, from supplementing the primary data. Data on the nature, extent and frequency of drought occurrences; population, area, districts and villages affected; historical records of droughts and famines; etc. were collected from available sources like the Rajasthan State Archives, Bikaner; State Administration Reports; Rajputana Gazetteers; Famine Reports; Rajasthan Directorate of Economics and Statistics; Famine and Relief Departments and various other State, district and tehsil offices.

The Nature of Dought Perception

Considering drought in relation to desertification, concepts and definitions have been evolved from time to time and from different angles. Closely tied to drought and its impacts on the physical and social environments has been the phenomenon known as desertification.

Where drought is largely a natural phenomenon, social responses and poor landuse practices greatly amplify its impacts. When severe drought is combined with land misuse the result is often permanent in land deterioration and loss of productivity. That is, desertification.

Desertification, in turn intensifies the effects of drought.

TABLE 3.1: **Sample villages covered, households enumerated and schedules filled in from different ecological tracts of arid zone of Rajasthan**

Ecological tract and village	District/ tehsil	Households enumerated	Schedules filled
I. *Predominantly pastoral* (165 mm mean annual district rainfall):			
1. Bandhewa	Jaisalmer/ Pokaran	294	29
2. Bharewala	-do-	133	13
3. Khetolai	-do-	160	16
4. Sankada	-do-	652	66
II. *Predominantly rainfed* (388 mm mean annual) district rainfall):			
1. Bhojas	Nagaur/Nagaur	152	15
2. Chhajoli	Nagaur/Jayal	202	20
3. Rohina	-do-	251	25
4. Rohini	Nagaur/Nagaur	296	30
III. *Predominantly rainfed with irrigation* (490 mm mean annual) district raintall):			
1. Dhabar	Pali/Pali	231	23
2. Dhurasni	Pali/Sojat	137	14
3. Jheentra	Pali/Pali	332	33
4. Sardar Samand	Pali/Sojat	250	25
Total		3090	309

Drought has many faces and means different things to different people. Some consider it a natural phenomenon due to climatic fluctuations while some consider it a man-induced bio-physical to socio-religious phenomenon inducing various problems. Socially, it may be equated with a state of functional and structural community disorganisation. Coupled with increasing human activity this results into social and economic degradation of the population and over-exploitation of the productivity of arid lands. Thus it is viewed as a pathological condition afflicting the Indian desert society. There is, therefore, a need too study man-environmental relationship. Such a relationship is revealed through existing knowledge of the rural folk viewing drought in relation to values and perceptions of individuals. The importance of perception lies in preparing the farmers socially and psychologically to face drought hazards and to evolve adjustment mechanisms.

Drought as a socially perceived phenomenon varies differently in different ecological tracts representing different cultures. The psychology of perception involves the whole field of attitudes and opinions of judgments and belief. Each individual's experience combine in a complex fashion to determine his reaction to a given stimulus situation viz. droughts. To the extent that certain classes of experiences are more likely to occur in some cultures than in others, differences in behaviour across cultures (different ecological tracts) including differences in perceptual tendencies, can be great enough even to surpass the ever present individual differences within cultural groupings.

The study reveals significant differences across cultures in susceptibility of several types of droughts, illusions; these differences are not genetical differences. They are differences produced by the same kinds of factors that are responsible for individual differences in illusion susceptibility, i.e. differences in experience. The findings reported, and the findings of others reviewed, point to the conclusion that to a substantial extent one learns to perceive; that inspite of the phenomenally absolute character of our perceptions, they are determined by perceptual inference habits and that various inference habits are differently liked in different ecological cultural societies. For all mankind, the basic process of perception is the same; only the contents differ and these differ only because they reflect different perceptual inference habits.

People's Perception of Drought

Local awareness of drought is as old as the Thar desert itself. Centuries old adaptation of the desert dwellers with the environment and local descriptions in the form of proverbs and folklores imbedded into the cultural framework of the desert society are some of the evidences available. Adjustments to the local climate took place through a drought induced learning process. Dependency of the livelihood on the local precipitation patterns is a factor inducing more accurate perceptions of the drought hazard. There is a growing awareness and appreciation of the desert environment with the public perception of the seriousness of problems associated with drought. Before analysing and assessing drought induced problems like desertification, it is essential to know people's perception of drought and desert.

Nature and extent of drought years perceived, its occurrences and factors responsible, years witnessed and indicators evolved for prediction of the nature of year, formed the core of drought perception. Differential pattern of habitation comprising of different communities, different occupations under different ecological patterns, revealed different perceptions of drought behaviour of the desert dwellers.

The study on the aspect of drought perception provides an approach to the ways in which the desert dwellers perceive and cope with their main problem, the recurring hazard of drought. Earlier various hypothesis set up for testing the following appear to have been substantiated. Perception of the drought hazard varied according to: (1) degree of aridity, (2) amount of drought experience, (3) personality differences, (4) age of the respondent, (5) educational level, and (6) type of occupation (Saarinen, 1966). In this study the strongest single relationship found was that the perception of drought hazard varied with aridity. As aridity increased so does the number of drought years (Table 3.2).

TABLE 3.2: **Drought problems experienced by the respondents of different ecological tracts in the arid zone of Rajasthan**

Ecological Tract	Respondents reporting problems				Total number
	Yes (No.)	% of total	No (No.)	% of total	
Pastoral	122	98	2	2	124
Rainfed	70	78	20	22	90
Irrigated	65	68	30	32	95
All tracts	257	83	52	17	309

Data revealed that the respondents, irrespective of ecological tract, had an appreciation of the problem of drought in the area. But the respondents of pastoral tract as compared to the rainfed and irrigated tracts had greater perception of the frequency of drought in the area. About 98 per cent of the respondents of pastoral tract as compared to 78 per cent, 68 per cent of the rainfed and irrigated tracts had reported frequent drought problem in their respective area.

Perception was further revealed in terms of Zamana — the local unit of measurement of mainly Kharif harvest (Table 3.3).

TABLE 3.3: Nature of extent of drought and good years perceived in terms of Zamana (1891 to 1984)

Nature of years with Zamana (%)	Number of years perceived			Mean of tracts
	Pastoral	Rainfed	irrigated	
Severe drought (< 10% Zamana)	8	5	5	6
Droughts (11-25% Zamana)	35	25	13	24
Mild drought (26-50% Zamana)	19	30	33	27
Normal year (51-75% Zamana)	11	16	20	16
Good year (76-100% Zamana)	21	18	23	21
Total (0-100% Zamana)	94	94	94	94

Out of 94 years (1891 to 1984) the number of drought and severe drought years perceived were found to be 43 in case of pastoral tract as compared to 30 and 18 in the rainfed and irrigated tracts respectively. Thus, severity of droughts perceived increased with the decrease in rainfall and vice-versa. Mild drought and normal years perceived increased with the increase in rainfall. The people had on an average perceived 6 severe drought years having less than 10 per cent Zamana; 24 drought years (11-25% Zamana); 27 mild drought years (26-50% Zamana); 16 normal years (51-75% Zamana) and 21 good years (76 — 100% Zamana).

Relation of drought perception was further revealed by drought frequency. Respondents were asked to report upon, how often drought occurs in the area (Table 3.4).

Drought though a common feature of the area, was reported to be most frequently occurring in the pastoral tract as compared to other tracts. In a pastoral tract 45 per cent of the total respondents reported occurrence of drought every year, while 30 per cent reported it in an alternate year and 25 per cent expressed after 2 years interval. In a rainfed tract drought occurrence was reported every year by 15 per cent, alternative year (20%), after 2 years (50%) and after 3 years (15%). In an irrigated tract the emphasis had shifted to a greater interval of drought occurrences. About half of the respondents had stated drought occurrence after 2 and 3 years interval. This shows that the respondents of pastoral tract had greater perception of drought because of greater amount of drought experience.

TABLE 3.4: **Occurrences of drought in the area**

Frequency of occurrence	Percentage of respondents			All tracts
	Pastoral	Rainfed	Irrigated	
Every year	45	15	–	20
Alternate year	30	20	6	19
After 2 years	25	50	48	41
After 3 years	—	15	46	20

Perception of the drought risk varied according to the type of occupation followed. The farmers (65%) of pastoral tract predominantly engaged in livestock rearing though dependent on rainfall are generally less perceptive of the drought risks. These livestock breeders perceive the risk somewhat less serious (54% as compared to 41 per cent reporting more serious), than those whose main emphasis is on agricultural only (63% respondents reported more serious than 30 per cent reporting less serious). Cultivators of rainfed and irrigated tracts were better perceptive of the drought risks. Their main occupation of cultivation is chiefly dependent upon rainfall and perceive more seriousness (64% and 54% as compared to 27% and 41% reporting less serious) of the consequences of drought.

Perception of Drought Problems

Desert rural-folk view drought from various angles. Their notions concerning the causes of drought and rainfall vary from climatic to bio-physical to socio-religious in nature. Highest percentage (77%) of responses were meteorological in the conventional sense citing climatic changes in wind direction, clouds, intensity and duration of storms and cold and hot seasons as the relevant causes. Biophysical reasons were reported by about half (49%) of the respondents. Almost non-existence of forests; devastation of natural vegetation by indiscriminate cutting for fuel, construction and field bunding purposes; traditional practices of overgrazing and frequent lopping trees by livestock raisers; and wind erosion causing shifting of dunes creating new desert land forms and land scapes on cultivated fields, village pastures and waste-barren lands were mentioned as the chief bio-physical factors of drought occurrences. Only one-third (33 per cent) of the responses revealed socio-religious causes such as supernatural beliefs, revenge by nature, bad luck, lack of religious faith, and immoral activities. All this conformed to the traditional pattern of reaction in a desert society.

Associated with these notions drought has further been viewed from multi-dimensional problems of the disorder in nature.

The problems involving depreciation of the socio-economic, bio-physical and cultural resource utilisation related in whole or in part to droughts accelerating desertification in the area. Such induced problems of socio-cultural, economic and bio-physical in nature were reported by 98 per cent of the inhabitants experiencing droughts in the areas. Of all the problems, socio-economic problems were the most predominant as revealed by more than four-fifths of the responses. While bio-physical and socio-cultural problems in nature were supported with 75 per cent and 60 per cent of the responses respectively.

Bio-physical problems reported were due to the fulfilment of the needs of the weaker segments constituting majority of the population. Local vegetation like trees and shrubs — *Khejri* (for local use and sale for fuel), *Banwarli* (for leather tanning industry), *Kumbhat* trees (for making wooden curd churner), *Phog* shrubs (for fuel and coal conversion); etc., were cut and sold chiefly by the non-agro-pastoral communities during the past droughts of 1965-66, 1968-69 and 1969-70 destroying 80 per cent of the total vegetation in the area. Wind erosion had affected village roads, houses, fields and Nadis (water ponds) in the village. The problems of blowing of fertile soil and covering of cultivated fields; silting of village wells, Nadis; were reported by a majority of the villagers. Huge livestock pressure and traditional desert life-styles associated with livestock rearing and personal needs were the major factors in upsetting land-livestock vegetation relationship. Though some of the best Sewam (local grass) grazing lands existed in the pastoral tract, yet improper control of the time and intensity of the grazing had devasted these rangelands by complete stripping of vegetation roots and all. These existing practices had made the land highly susceptible to erosion.

TABLE 3.5: Respondents reported problems during drought periods (multiresponses)

Problems	Households	
	No.	%
Socio-cultural (Migration; occupational diversification; Shifting settlements; disintegration of joint households; social losses; changes in social values; non-functioning of the socio-cultural institutions; disturbance in Jajman-Kamin relationship)	185	60
Socio-Economic (Loss of crop and livestock production; pests, diseases and poor crop growth; livestock mortality; setback to occupational caste's economy; distress sale of land, livestock personal assets; increased indebtedness; scarcity of water, grass fodder)	253	82
Bio-physical (Removal of vegetation; overgrazing; wind erosion; predominant barren lands; shifting dunes; silting of village Nadis-water ponds; over-exploitation of ground water)	232	75

People's Perception of Desertification

In arid zone of Rajasthan drought inducing desertification, is perceived by the desert dwellers, as a multi-dimensional phenomenon. It is chiefly perceived in the form of changes in climatological, bio-physical, socio-cultural and economic parameters undergoing disorders in nature. It is also viewed as a product of man-land-livestock-vegetation relationship under ecological stress forming components of desertification processes in the area. The processes involve gradual deterioration of land facing landscape changes; changes in eco-climatological and socio-economic parameters accelerating desertic conditions (Table 3.6).

There is a growing awareness and appreciation of the

intensification of the desert environment developing in arid zone of Rajasthan. This has been reflected in the shift towards desert land deterioration and in the people's perception of the intensification of problems associated with desertification. Although people's consciousness and appreciation of the changes in the desert landscape appear to be a relatively recent phenomenon, the Thar desert has long been perceived a natural gift *Dhora Dharti* (land of sand dunes). The sunny desert climate is considered an asset for rearing large herds of animals viz. cattle, sheep, goats and camels.

TABLE 3.6: **Nature and extent of the parameters of desertification perceived and changes noticed during the last 30 years**

Parameters	Percentage of the respondents reported		
	Increase	Decrease	No change
1	2	3	4
1. *Climatological*:			
Rainfall	20	45	35
Dust storms	5	40	55
Scarcity periods	21	16	63
Severe droughts	45	30	25
Drought frequency	40	32	28
Aridity	48	12	40
Water table	–	78	22
Over-exploitation of ground water	65	20	15
Deterioration in quality of water	43	21	36
2. *Ecological*:			
Vegetation cover-extent	–	85	15
Deterioration in quality of vegetation	64	12	24
Deterioration of grazing lands	73	–	27
Undesirable weeds	68	10	22
Deforestation/cutting of vegetation	79	10	11
Overcrowded grazing lands	84	6	10
Water logging of cultivable lands	45	–	55

(*Contd.*)

TABLE 3.6: (*Contd.*)

1	2	3		4
	Salinisation of irrigated lands	55	5	40
	Degradation of rainfed croplands	62	6	32
3.	*Socio-economic*:			
	Human population	100	–	–
	Livestock population	95	–	5
	Migration for livelihood	42	18	40
	Crop yield/unit area	7	73	20
	Livestock production	6	78	16
	Land-livestock carrying capacity	4	85	11
	Livestock migration	52	25	23
	Agro-livestock occupation	56	12	32
	Man-land ratio	–	77	23
	Over cultivation	54	21	25
	Fallowing of lands	4	78	18
	Land productivity	21	66	13
	Pressure on land	74	20	6
	Small/marginal farmers	45	22	33
	Ploughing marginal and submarginal lands	67	11	22
4.	*Land scape changes*:			
	Sand dune movement	32	15	53
	Hummocks on the land	7	23	70
	Soil erosion	43	15	42
	Sand blowing on fields	53	13	34

Source: Field data collected and analysed.

Aridity, less rainfall, vast uncultivated and wastelands and very nutritive native grasslands like 'Sewan' known as the king of the grasses are considered the most favourable factors of the desert climate for rearing livestock. Open spaces and dunes, interdunal plains, cooled and pleasant nights during summer were other natural features seen as the advantages of the desert region. Desert appeared as an advantage in only one per cent of the sample interviews. Despite frequent droughts and famines; scarcity of grain, water and fodder; very hard socio-ecological conditions; and extreme summers and winters; the desert dwellers do not want to depopulate this area permanently.

Of the various parameters of desertification, climatological followed by ecological, landscape and socio-economic were the most predominantly perceived parameters having undergone changes during the last thirty years. Among climatological, decrease in water table and rainfall (reported by 70 per cent and 45 per cent of the respective respondents); and increases in over-exploitation of ground water, aridity, severity of droughts and its frequency (65 per cent, 48 per cent and 40 per cent respectively) were the most predominant. Ecological changes reported were chiefly decrease in vegetation cover and its extent (85%), followed by increases in over-crowded grazing lands (84%), deforestation and cutting of vegetation (79%), deterioration of grazing lands (73%), undesirable weeds (60%) and quality of vegetation (65%). Landscape changes were reported to be increase in sand blowing on fields (53%), soil erosion (43%), sand dune movement (32%) with an insignificant decrease in them. A majority of the respondents reported no change in landscape parameters.

Socio-economic parameters accelerating the process of desertification were reported to be increase in human and livestock population reported by 100 per cent and 95 per cent respondents respectively. This followed by an increases in pressure on land, ploughing of marginal and sub-marginal lands, over cultivation, livestock migration and increase in small and marginal farmers reported by 74, 67, 54, 52 and 45 per cent respondents respectively. Decreases in land-livestock carrying capacity, fallowing of lands, livestock production, man-land ratio, crop yield/unit area were reported by 85, 78, 77 and 73 per cent respondents respectively.

Drought Recollection

The correlation between the farmers' recollection of past harvests and the record of actual rainfall in western Rajasthan is very close. Living memory of the nature of previous years extended as far back as 1891 with substantial agreement. Each year was remembered in terms of zamana. Full zamana is the level of Kharif crop production expected, given sufficient water. The classification of the years from 1891 to 1984 is given in terms of remembered zamana in Tables 3.7(a) and 3.7(b) (Informants actually talked in terms of annas — the old one-sixteenth division of the rupee. Annas have been translated into percentages for convenience). Zamana perceived was closely associated with rainfall and aridity of the area. The mean annual zamana perceived and remembered in pastoral tract varied from minimum 0 per cent (severe drought year) to maximum 55 per cent (normal year) and 77 per cent (Good year) with an overall 34 per cent zamana at 164 mm rainfall. In a rainfed tract of mean annual rainfall of 388 mm. the mean annual zamana perceived varied from minimum 5 per cent to maximum 60 and 80 per cent with an overall mean zamana of 52 per cent. In an irrigated tract of comparatively higher rainfall (490 mm) and lesser aridity the zamana perceived varied from 60 to 65 to 100 per cent respectively with an overall mean zamana of 61 per cent [Table 3.7(b)]. On an average a decade witnessed 6 severe drought years, 2 good years and hardly 2 average or "normal" years. It is worth noting that the high average rainfall for years of severe drought in Table 3.7(b) shows that zamana correlates not with amount of rainfall alone but with a combination of amount, timing and intensity. In this connection it is worth remembering that we have no direct record of one particularly significant indicator of a bad year – the extent of damage from flooding.

Recollection did of course vary. Generally, food producers — farmers and pastoralists — scored better than members of non-food-producing caste groups, though no significant variation was evident in relation to mild drought or "average" years. Much of the variation altogether arose from factors such an uneven spatial distribution of rainfall, landholdings, fallow and herds.

TABLE 3.7: Correlation of rainfall records and recollection of zamana

(a) by year

Year	Annual Rainfall (mm)			Zamana (%)		
	Jaisalmer	Nagaur	Pali	Jaisalmer	Nagaur	Pali
1	2	3	4	5	6	7
1891	115	228	278	26-50	26-50	26-50
1892	278	188	438	76-100	11-25	51-75
1893	380	513	438	76-100	76-100	51-75
1894	290	447	593	76-100	76-100	76-100
1895	75	298	193	11-25	26-50	11-25
1896	90	233	497	11-25	26-50	51-75
1897	278	360	258	76-100	51-75	26-50
1898	98	280	258	11-25	26-50	26-50
1899	8	60	18	0-10	0-10	0-10

(Contd.)

TABLE 3.7 *(Contd.)*

1	2	3	4	5	6	7
1900	125	303	528	51-75	76-100	76-100
1901	93	203	128	0-10	11-25	0-10
1902	130	190	315	11-25	11-25	26-50
1903	105	455	428	11-25	26-50	26-50
1904	53	325	298	11-25	51-75	26-50
1905	90	78	83	11-25	0-10	0-10
1906	313	270	523	11-25	26-50	76-100
1907	238	438	608	11-25	76-100	76-100
1908	425	653	792	76-100	76-100	76-100
1909	213	498	438	51-75	76-100	51-75
1910	163	335	768	11-25	51-75	76-100
1911	65	228	248	11-25	26-50	26-50
1912	190	303	360	51-75	51-75	51-75
1913	130	130	260	11-25	11-25	26-50
1914	258	240	278	11-25	11-25	26-50
1915	103	318	262	11-25	11-25	11-25
1916	330	455	722	76-100	76-100	76-100
1917	565	793	930	11-25	26-50	26-50
1918	0	63	78	0-10	0-10	0-10
1919	98	233	172	11-25	26-50	11-25
1920	275	170	352	26-50	11-25	51-75
1921	68	188	275	11-25	11-25	26-50
1922	123	308	208	26-50	11-25	11-25
1923	200	393	285	51-75	26-50	26-50
1924	185	195	220	51-75	11-25	11-25
1925	73	175	233	11-25	11-25	11-25
1926	315	300	760	76-100	51-75	76-100
1927	173	308	495	26-50	26-50	26-50
1928	98	463	448	11-25	11-25	26-50
1929	193	420	645	51-75	51-75	76-100
1930	70	323	298	11-25	11-25	11-25
1931	270	260	558	76-100	26-50	76-100
1932	168	178	383	51-75	11-25	51-75
1933	260	420	348	76-100	76-100	76-100
1934	135	373	415	11-25	26-50	51-75
1935	170	225	363	11-25	11-25	26-50
1936	153	335	250	11-25	11-25	11-25
1937	243	330	283	26-50	26-50	26-50
1938	108	162	498	26-50	11-25	51-75
1939	40	110	275	0-10	11-25	26-50
1940	285	448	473	76-100	76-100	76-100
1941	113	148	255	11-25	11-25	26-50
1942	168	508	390	76-100	76-100	51-75
1943	90	273	478	11-25	26-50	51-75

(Contd.)

TABLE 3.7 *(Contd.)*

1	2	3	4	5	6	7
1944	570	378	862	51-75	51-75	76-100
1945	8	340	442	0-10	26-50	51-75
1946	20	80	260	0-10	0-10	26-50
1947	210	260	467	26-50	26-50	51-75
1948	78	NA	NA	11-25	0-10	11-25
1949	105	338	285	26-50	51-75	26-50
1950	115	422	568	76-100	76-100	76-100
1951	108	138	403	11-25	11-25	51-75
1952	145	133	515	11-25	11-25	26-50
1953	178	803	665	26-100	76-100	76-100
1954	273	273	390	51-75	26-50	26-50
1955	433	390	718	26-50	51-75	76-100
1956	201	382	510	76-100	51-75	76-100
1957	188	395	396	26-50	51-75	26-50
1958	151	268	265	76-100	51-75	26-50
1959	232	324	576	51-75	51-75	51-75
1960	136	192	287	11-25	11-25	26-50
1961	532	297	388	76-100	51-75	51-75
1962	256	283	285	26-50	26-50	26-50
1963	67	138	350	11-25	11-25	26-50
1964	128	588	470	26-50	76-100	51-75
1965	99	221	485	11-25	26-50	51-75
1966	102	137	505	26-50	11-25	26-50
1967	167	354	783	51-75	51-75	76-100
1968	83	285	293	11-25	26-50	11-25
1969	25	178	101	0-10	11-25	0-10
1970	258	571	600	76-100	76-100	76-100
1971	114	467	248	11-25	26-50	11-25
1972	68	331	239	11-25	26-50	11-25
1973	335	798	778	76-100	76-100	76-100
1974	31	295	172	0-10	26-50	11-25
1975	414	1223	775	76-100	51-76	76-100
1976	369	489	368	26-50	26-50	26-50
1977	131	503	544	26-50	26-50	51-75
1978	215	450	541	76-100	76-100	76-100
1979	166	246	769	26-50	26-50	51-75
1980	112	228	305	26-50	26-50	26-50
1981	147	391	256	26-50	51-75	26-50
1982	132	389	378	26-50	26-50	26-50
1983	251	635	704	76-100	76-100	76-100
1984	112	207	322	26-50	26-50	26-50

(b-i) by % Zamana in a pastoral tract

% Zamana	Rainfall (mm)	Mean Zamana	Years
0 – 10% (Severe drought)	0 – 50	0%	1899, 1901, 1918, 1939, 1945, 1946, 1969, 1974
11 – 25% (Drought)	51 -100	12%	1895, 1896, 1898, 1902, 1903, 1904, 1905, 1906, 1907, 1910, 1911, 1913, 1914, 1915, 1917, 1919, 1921, 1925, 1928, 1930, 1934, 1935, 1936, 1941, 1943, 1948, 1949, 1951, 1952, 1960, 1963, 1965, 1968, 1971, 1972,
26 – 50% (Mild drought)	101 -150	28%	1891, 1920, 1922, 1927, 1937, 1938, 1947, 1955, 1957, 1962, 1964, 1966, 1976, 1977, 1979, 1980, 1981, 1982, 1984
51 – 75% (Average)	151 -200	55%	1900, 1909, 1912, 1923, 1924, 1929, 1932, 1944, 1954, 1959, 1967
76 – 100% (Good)	201 +	77%	1892, 1893, 1894, 1897, 1908, 1916, 1926, 1931, 1933, 1940, 1942, 1950, 1953, 1956, 1958, 1961, 1970, 1973, 1975, 1978, 1983,
Overall mean	164	34%	1891 -1984 (94 Years)

(b-ii) by % Zamana in a Rainfed tract

% Zamana	Rainfall (mm)	Mean Zamana	Years
0 – 10% (Severe drought)	0 – 100	5%	1899, 1905, 1918, 1946, 1948
11 – 25% (Drought)	101 – 200	25%	1892, 1901, 1902, 1913, 1914, 1915, 1920, 1921, 1922, 1924, 1925, 1928, 1930, 1932, 1935, 1936, 1938, 1939, 1941, 1951, 1952, 1960, 1963, 1966, 1969
26 – 50% (Mild Drought)	201 – 300	42%	1891, 1895, 1896, 1898, 1903, 1906, 1911, 1917, 1919, 1923, 1927, 1931, 1934, 1937, 1943, 1945, 1947, 1954, 1962, 1965, 1968, 1971, 1972, 1974, 1976, 1977, 1979, 1980, 1982, 1984
51 – 75% (Average)	301 – 400	60%	1897, 1904, 1910, 1912, 1929, 1944, 1949, 1955, 1956, 1957, 1958, 1959, 1961, 1967, 1975, 1981
76 – 100% (Good)	401 +	80%	1893, 1894, 1900, 1907, 1908, 1909, 1916, 1926, 1933, 1940, 1942, 1950, 1953, 1964, 1970, 1973, 1978, 1983
Overall mean	388	52%	1891 -1984 (94 Years).

(b-iii) by % Zamana in an Irrigated-cum-rainfed tract

% Zamana	Rainfall (mm)	Mean Zamana	Years
0 – 10% (Severe drought)	0 – 150	10%	1899, 1901, 1905, 1918, 1969
11 – 25% (Drought)	151 – 250	25%	1895, 1915, 1919, 1922, 1924, 1925, 1930, 1936, 1948, 1968, 1971, 1972, 1974
26 – 50% (Mild Drought)	251 – 350	48%	1891, 1897, 1898, 1902, 1903, 1904, 1911, 1913, 1914, 1917, 1921, 1923, 1927, 1928, 1935, 1937, 1939, 1941, 1946, 1949, 1952, 1954, 1957, 1958, 1960, 1962, 1963, 1966, 1976, 1980, 1981, 1982, 1984
51 – 75% (Average)	351 – 500	65%	1892, 1893, 1896, 1909, 1912, 1920, 1932, 1934, 1938, 1942, 1943, 1945, 1947, 1951, 1959, 1961, 1964, 1965, 1977, 1979
76 – 100% (Good)	500 +	100%	1894, 1900, 1906, 1907, 1908, 1910, 1916, 1926, 1929, 1931, 1933, 1940, 1944, 1950, 1953, 1955, 1956, 1967, 1970, 1973, 1975, 1978, 1983
Overall mean	490	61%	1981 -1984 (94 Years)

The basis of the memory of drought years can be related to various factors. Earlier disastrous years — 1812-13, 1868-69, 1877-78, 1891-92, were cited by a few old men on the basis of inherited information. Some severe droughts and famines, like 1899, 1901, 1918, 1939,1960, 1968, 1969, etc. were remembered with accuracy due to reinforcement or feedback from documentation of migration, mortality, desertion of villages, or scarcity of grain, water, and fodder in sources such as the Rajputana Gazetteers and State Administration and Famine Commission Reports. There was also a clear folk memory of good years in 1900, 1908, 1916, 1926, 1933, 1940, 1944, 1954, 1955, 1961 1970 etc. For example, everyone knew that in 1940 the rains had continued till the month of Asoj (September-October) and the month harvest had excelled all records; that a record bajra crop had been recorded in 1961; and that in 1921 the bajra crop had spoiled as a result of additional late rainfall but moth (*phaseolus aconitifolius*) production had excelled.

There was an overall tendency to exaggerate the severity of the droughts as they receded into the past, especially on the part of food-producers. Moreover, earlier droughts were associated with not only less rainfall, but more intense storms, more soil erosion, shifting dunes, more crop losses, migration, livestock mortality, desertion of villages, as well as the lack of motorised transport, drinking water, drought-resistant crop variations. On the other hand, there was a general awareness that population levels were lower, that there was less crop disease, that joint households were more cohesive and resistant to centrifugal forces, that vegetation was denser, and that land was more often fallowed, so that the effects of drought were not so immediately disastrous, despite their intensity, because of the greater margin between exploitation levels and carrying capacity, and possibly also because of greater social cohesiveness.

Discussion of the causes of drought in Rajasthan ranges over a variety of factors from climatic and biophysical to social and religious. Although most informants related the occurrence of drought primarily to physical factors such as changes in wind direction, cloud cover, frequency and intensity of storms, and the overall quality of cold and hot seasons, many emphasised the role of bio-physical factors such as decline in forest and vegetation cover, increased wind erosion and shifting dunes and silting of village ponds. There was a general consciousness that human activities lay behind these natural factors, particularly indiscriminate cutting for fuel and construction, overgrazing and lopping of trees for fodder, but considerable emphasis was also given to supernatural and moral factors such as luck, apathy immorality and the revenge of nature.

In discussion of the affects of drought most people first cited economic disorder; failure of crops, loss of livestock, reduction in the value of assets, and forced sale. But there was only slightly less emphasis on social disorder; migration, the failure of institutions, disintegration of social groups and of households.

Drought continues to be seen as a form of instability or disorder in both the natural and the moral world. But the context of instability has changed. Pastoralists may migrate long distances in search of land for grazing in a drought year, and although earlier there was enough vacant territory to absorb them, now farmers complain that they not only exhaust limited water and grass resources, but also destroy the standing crops, disrupt soil conservation measures and create conflict and tension.

The problems caused by the instability have become more complex not only because of the increased density of the population but because of recent socio-economic changes. For example, symbiosis of food-producing and occupational caste groups has been disturbed.

Members of occupational castes such as potters, leather-workers, oil-pressure, and retailers, have now acquired land and become partially dependent on cultivation and have thus lost flexibility while individual craftsmen may have gained flexibility through economic diversification, in many cases the crafts have gone into decline. Joint households have broken down into nuclear households causing greater pressure on scarce resources by division and fragmentation of property. Participation in social events has narrowed from the community at large to the smaller extended kinship group. There has been general decrease in family solidarity and increase in conflict, tension and insecurity. During the drought of 1968-1969 people sold land, livestock, ornaments, houses and household articles, and the *jajmani* or *aat* system of payment in kind or cash between caste groups for services rendered was disrupted with the result that the existing trend of socio-economic change was accelerated.

As the social and economic context of drought has changed, so has its meaning. The change in meaning is more difficult to reconstruct. But it is worth noticing that many biophysical effects reported from recent droughts derive from the responses of the weaker segments of the population. For example, particular trees and shrubs (Khejri-*Prosopis cineraria* — for fuel; phog-*Calligonum poligonoides* — for charcoal; banwarli for tanning; *Kumbhat* for churns) were cut and sold in larger quantities.

Although the context of the experience and recollection of drought has been changing, the drought lore followed by the desert dwellers may not be changing in the same way. It is likely, however, that its use is changing. By analogy with equivalent processes elsewhere it might be expected that this lore will gradually be relegated to the epistomological status of superstition, but such a change in the form of knowledge does not prove a change in the manner of thinking. The study of this type of thinking can be expected to lead to a better understanding of non-scientific thought generally (Spooner and Mann, 1982).

Summary and Conclusion

Perception of drought and desertification has brought out people's awareness and understanding of the causes and processes of the environmental phenomenon. The paper investigates the memory of drought over a period extending back to the limits of living memory in the villages of western Rajasthan. This adds to our understanding of how memory works in a cultural system which includes the synoptic perception of past droughts as all-encompassing phenomena — a perception on which is based the expectation and prediction of drought to come. It represents memory as the adhoc ordering of recollections and their commonsense rationalisation.

Drought has been considered a natural phenomenon due to climatic fluctuations and a man-induced bio-physical to socio-religious phenomenon inducing various problems. Socially, it is equated with a state of functional and structural community disorganisation. Coupled with increasing human activity this results into social and economic degradation of the population and over-exploitation of the productivity of arid lands.

Differential pattern of habitation comprising of different communities, different occupations under different ecological patterns, revealed different perceptions of drought behaviour of the desert dwellers. Perception of the drought hazard varied according to degree of aridity, amount of drought experience, age of the respondent and type of occupation. The strongest single, relationship found was that the perception of drought hazard varied with aridity. As aridity increases so does the number of drought years. Drought was reported to be most frequently occurring in the pastoral tract as compared to other tracts. The respondents of pastoral tract had greater perception of drought because of greater amount of drought experience.

Respondent's notions concerning the causes of drought vary from climatic to bio-physical to socio-religious in nature. Most informants related the occurrence of droughts primarily to physical factors such as changes in wind direction, cloud cover, frequency and intensity of storms; and the overall quality of cold and hot seasons many emphasised the role of bio-physical factors such as decline in forest and vegetation cover, increased wind erosion and shifting dunes and silting of village ponds. There was a general consciousness that human activities lay behind these natural factors, particularly indiscriminate cutting for fuel and construction, overgrazing and lopping of trees for fodder, but considerable emphasis was also given to supernatural and moral factors.

The correlation between the farmers recollection of past harvests and the record of actual rainfall in western Rajasthan is very close. Each year was remembered in terms of Zamana (level of Kharif crop production expected). Zamana perceived was closely associated with rainfall and aridity of the area. On an average a decade witnessed 6 severe drought years, 2 good years and hardly 2 average or normal years.

Earlier droughts were associated with not only less rainfall, but more intense storms, more soil erosion, shifting dunes, more crop losses, migration, livestock mortality, desertion of villages, as well as the lack of motorised transport, drinking water, and drought-resistant crop varieties. But there was a general awareness that population levels were lower, there was less crop disease, joint households were more cohesive and resistant to centrifugal forces, vegetation was denser, and that land was more often fallowed, so that the effects of droughts were not so immediately disastrous, despite their intensity, because of the greater margin between exploitation levels and carrying capacity, and possibly also because of greater social cohesiveness.

Drought inducing desertification is chiefly perceived in the form of changes in climatological, bio-physical, socio-cultural and economic parameters undergoing disorders in nature. Socio-economic parameters accelerating desertification were increased in human and livestock population bringing pressure on land, ploughing of marginal and sub-marginal lands, over-cultivation, livestock migration and increase in small and marginal farmers. Decrease in land-livestock carrying capacity, fallowing of lands, livestock production, man-land ratio, crop yield/unit area were the consequences reported by the respondents.

Drought thus continues to be seen as a form of instability or disorder in both the natural and the moral worlds. The problems caused by the instability have become more complex, not only because of the increased density of the population but because of recent socio-economic changes.

Although the context of the experience and recollection of drought has been changing, the drought lore is not changing in the same way. It is likely, however, that its use is changing. By analogy with equivalent processes elsewhere it might be expected that this lore will gradually be relegated to the epistemological status of superstition, but such a change in the form of knowledge does not prove a change in the manner of thinking. The study of this type of thinking can be expected to lead to a better understanding of non-scientific thought generally. However, the value of this material will serve a key to understanding the coping mechanisms and potential reactions of a rural population.

REFERENCES

Bharara, L. P., 1978 : "Social Indicators of Drought Prediction in a Village Community of Rajasthan Desert", presented at *International Symposium-Post Plenary Session-Anthropology and Desertification, CAZRI*, Jodhpur.

———, 1980 : "Social Aspects of Drought Perception in Arid Zone of Rajasthan", *Annals of Arid Zone*. 19. (1 & 2) : pp. 154-157.

———, 1982 : "Notes on the Experience of Drought Perception, Recollection and Prediction", in Spooner B. and Mann, H.S., (editors), *Desertification and Development : Dryland Ecology in Social Perspective*, pp. 351-361, Academic Press, London.

———, 1986 : "Concept of Drought and Famine viewed from Sociological Angle", paper presented at *XI World Congress of Sociology*, New Delhi, August, pp. 18-22.

Brookfield, H. C., 1969 : "On the Environment as Perceived", In : *Progress in Geography*, Vol. I.C. Board, R.J. Chorley, P. Haggett, D.R. Stoddart (Eds), pp. 51-80, Edward Arnold, London.

Burton, I., Kates, R. W., 1964 : "The Perception of Natural Hazards in Resources Management", *Natural Resources Journal*, 3. pp. 412-441.

CAZRI, 1979 : *25 Years of Arid Zone Research 1952-77*, ICAR, CAZRI, Jodhpur

Craik, K. H., 1970 : "Environmental Psychology", *New Directions, in Psychology*, Vol. 4. Holt, Rinehart and Winston, New York.

DST, 1981 : "Report of the Working Group on Arid Zone Research", Govt. of India, p. 181.

Goodey, B., 1971 : *Perception of the Environment*, Occasional Paper 17, Centre for Urban and Regional Studies, University of Birmingham, Birmingham.

Krishnan, A., 1968 : "Distribution of Arid Areas in India", *Proc. Symp. on Arid Zone*, 21st International-Geog. Cong. Jodhpur, pp. 11-19.

Lowenthal, D., 1967 : *Environmental Perception and Behaviour*, Department of Geography Research Paper No. 109, University of Chicago, Chicago.

Malhotra, S. P., and Bharara, L. P., 1968 : "Population, Resources and Food Situation in Various Tracts within the Arid Zone of Rajasthan", *Indian Sociological Bulletin*, VI (4). pp. 213-221.

O'Riordan, T. O., 1971 : *Perspectives on Resource Management*, Pion, London.

Proshansky, N. M. Ittelson, W.H. and Rivlin, L.G., 1970 : *Environmental Psychology; Man and His Physical Setting*, Holt, Rinohart and Winston, New York.

Saarinen, T. F., 1966 : *The Perception of the Drought Hazard on the Great Plains*, Department of Geography, Research Paper No. 106, University of Chicago, Dept. of Geography, Chicago.

———, 1976 : *Environmental Planning: Perception and Behaviour*, Housing Mifflin, Boston.

Sewell, W. R. D., Burton, I (Eds.), 1972 : *Perceptions and Attitudes in Resources Management*, Information Canada, Ottawa.

Spooner, B. and Mann, H. S., 1982 : *Desertification and Development : Dryland Ecology in Social Perspective*, (Eds.) Academic Press, London.

Whyte, A. V. T., 1977 : *Guidelines for Field Studies in Environmental Perception*, UNESCO, Paris.

4

DIFFUSION, DEFORMATION OR EPIGENESIS? GROWTH PROCESSES OF COLONIA AGRICOLA IN ARGENTINA

ARTHUR S. MORRIS

A PRINCIPAL direction in research on spatial growth processes is that of diffusion. This approach is peculiarly applicable to the study of growth for single innovations over short periods of time. Its applicability to more complex growth processes is less clear. In the case of the planned agricultural colonies of the 19th century Argentina, the *colonia agricola* that spread was not a single object but a set of ideas and technological innovations, and there were numerous changes in the environment of the innovation set and in its content as a set, so that a straightforward diffusion model, thought it yields results, is not really appropriate, and the researcher is forced to turn to more loosely defined models of growth, involving also an element of differentiation as well as sheer increase in size or area covered.

First we may ask what was the character of this innovation, the *Colonia agricola*? It was not a simple innovation in the nature of those studied by Hagerstrand and others using simulation models. Items such as cars, television sets, gas lighting, hybrid corn are essentially unitary entities, and they have further advantage from the point of view of modelling, that their character remains essentially unchanged through the period of diffusion, so that concepts such as the relevant target population, the channels of communication and the effect of distance are parameters which are stable and well-known over the space and time dimensions of the study. In the case of the *Colonia agricola*, however, the object of study is not a single phenomenon but a colonising movement of variable character, whose units are pre-surveyed blocks of land, taken generally out of wasteland or great cattle ranches (*estancias*) and divided up into small lots suitable for intensive agriculture. This division is done under the administrative control of a private coloniser-agent, usually with governmental backing. We may in fact list the essential features as they were in the 1850s, and then compare them to the later period of the 1880s and 1890s. For the early period, basic features are :

1. Small farm size, usually 34 hectares, an innovation among the great *estancias* of Argentina.
2. Following from the small size, intensive landuse, usually for grain production, with a moderate mixture of crops and animals enforced by isolation and difficulties in transport for early colonies.
3. Owner-operator units of production, normally family farms.
4. State or province support, usually of an indirect form.
5. A planned and ordered layout of the *colonia*, with surveying of roads, from plots, public building sites, all completed before settlement.
6. Emphasis on attracting immigrant European farmers.

Diffusion

It would not be difficult to treat the *colonia agricola* as a typical diffusion. In the time dimension, it had a typical logistic curve type of acceptance pattern as it spread over the Pampas, exemplified by the data from Entre Rios Province (Fig. 4.1)

In spatial terms it is also a spatial diffusion process of unexceptionable character, dominated by a contagious spread away from centres in central and southern Santa Fe, around the first colony at Esperanza near Santa Fe city, and along the Centre Argentine railway from Rosario to Cordoba. Measures based on a stochastic or deterministic simulation would no doubt demonstrate this contagious growth pattern, but it is in any case directly apparent from the map (Fig. 4.2).

But as the spread takes place there is also a deformation of the original characteristics of the *colonia* until it ends as something quite different in content from the original. In place of the list of original characteristics above, there are :

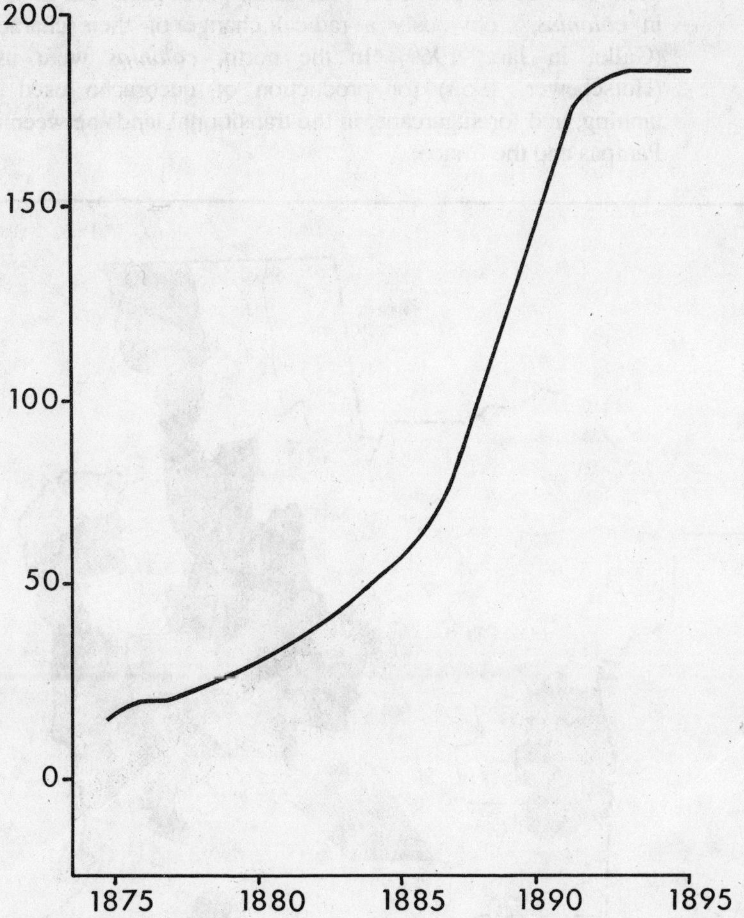

FIG 4.1: Rear of *colonias agricolas* in the province of Intre Los between 1875 and 1895.

1. Relatively large farms, consisting even by 1871 (Wilcken, 1873) of two, three or four lots of 34 hectares in place of one, and some being much larger.
2. In place of intensive grain production, many farms changed to a dairy farming system before the turn of the century, in the original areas of central Santa Fe. On the expanding edges of the *colonia* land in southernmost Santa Fe, and also in the

north-west of the province, beef cattle production was adopted in *colonias,* obviously a radical change of their character (Gallo, in Jara, 1969). In the north, *colonias* were used (Hotschewer, 1953) for production of quebracho used for tanning, and for sugarcane, in the transitional lands between the Pampas and the Chaco.

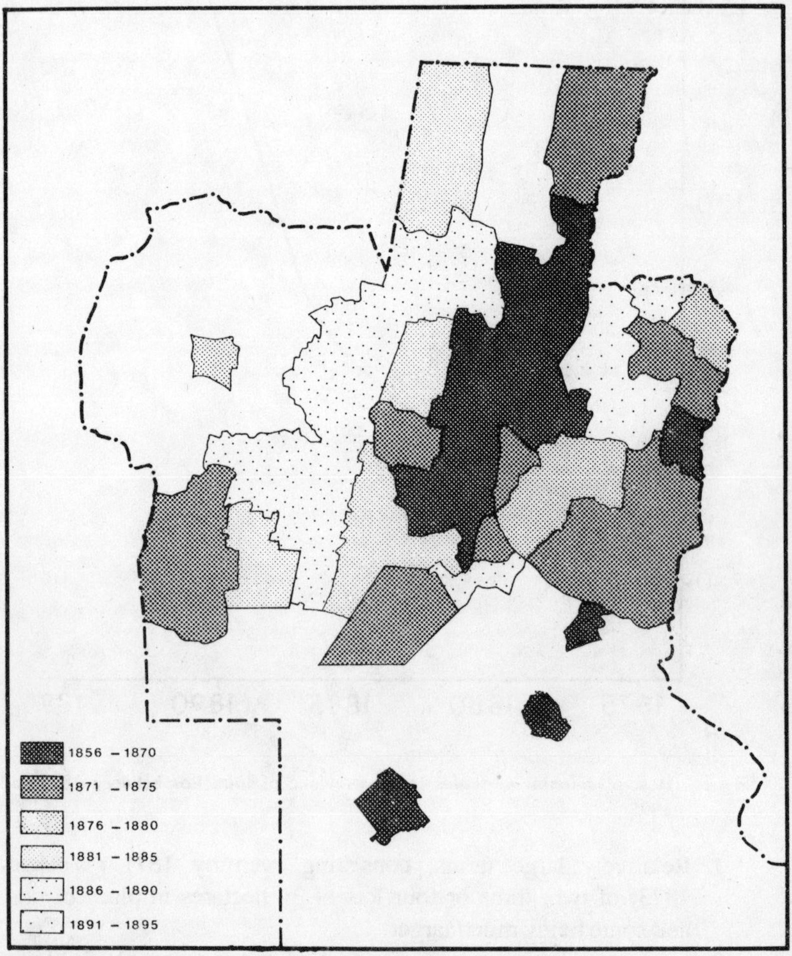

FIG. 4.2: Spatial growth pattern of the *colonia agricola.* The shadings show the time at which each department first had a *colonia.*

3. In place of owner-operator farms, tenant farms, even sub-tenancies, starting from the earliest colonies but penetrating

much more deeply with time, achieved a position of dominance in the 1890s and 1900s (Sternberg, 1972, vide especially the series, of maps). This fundamental change implied also a death blow to the idea that the *colonias* would provide a stable, hardworking, landowning and prosperous rural population; tenant farmers had little hold on or interest in the land, and finally left it in massive numbers after 1930.

4. In place of the State-supported schemes and even one or two attempts at direct State creation of *colonias*, later *colonias* were often totally private colonisation schemes, as *estancieros* saw an advantage in attracting small farmers to their land. This change brought with it the turn from social and political considerations in settlements — the reasons for State encouragement are discussed in Sternberg (1972) and in Jefferson (1926)—to largely economic motivation in the schemes of private ranchers for settlement on their land. In the last stages of colonisation, in La Pampa province on the western arid fringe of the Pampas (Gaignard, 1967) the colonisation movement was taken over by large companies with a purely commercial character, the main concern of administrators being the regular payment of rent or of the mortgage taken up by the small farmer. Social concern declined to the vanishing point.
5. A characteristic which did not change fundamentally was the pre-settlement survey. *Colonias* remained as planned settlements, in the narrow sense of surveying. Even in the plan there were some variations, away from the standard of single isolated farmsteads on each farm plot, which some writers have emphasised in comparing the various colonisation schemes. Kaerger (1901) in his detailed description of colonisation, especially of the German efforts in Argentina, Uruguay and Paraguay, writes of the efforts to establish a village form of settlement, with communally held land, local-self government and other features dating back to medieval European experience, on the part of the Volga German immigrants brought to Argentina when they lost their privileged position in Russia in 1880. Village settlements were opposed by local authorities and also by the central government, because they allowed foreign communities to remain too much together, and were also economically unjustifiable on the open Pampas territory with large farm sizes. Minor changes were allowed,

such as flexibility in the number of plots allocated to any family by leaving spare plots alongside each initial one, elimination of common land which was tried at first and found unworkable, and simplification of an excessively fine-meshed road network. But the basic pattern of farm and farmstead layout was retained.

6. Immigrant farmers continued to form a limited part of the human base for the colonies, especially those of an "ethnic" character, dominated by Jews and other European minority groups, but in general late colonies tended to admit only farmers with experience, thus effectively eliminating many immigrants as potential colonists; the farmers of these later settlements were often men who had moved on from the earlier colonies, with hard-won experience of grain-farming in the Pampas and, therefore, more suitable candidates for the land.

Deformation

What caused the changed content of the innovation as it spread out from the early centres? A first approximation might be to say that the decision-makers achieve an even closer approach to economic rationality as they become more experienced with the innovation, and that the changes are moves through adaptation or elimination in a kind of economic Darwinism, as expressed by White and Watts (1977). Another kind of approach is to say that over a period of time as that of the *colonia agricola's* spread, there are bound to be variations in the environment of the innovation, and technological changes will especially affect the kind of agricultural system employed. An alternative on this kind of theme is to say that the innovation also spreads over a wide geographical range of environments and, therefore, must adapt as it moves. But the argument takes a new turn if we acknowledge that the environment dos not only present a changing face as the innovation spreads through it in space and time; the environment may be active rather than passive, it may present counter-innovations, and we may interpret the changes as the result of a clash between two spreading innovation groups. This in turn leads to a consideration of the nature of the set or group involved in innovation, and some comments will be made, on the kinds of classification of change processes which are needed to accommodate such phenomena. A brief review of the major categories of explanation may first be made.

Perception and Decision-Making

One way of looking at the changes is to view them as progressive increases in rationality due to better information and perception levels as time went on. Changes in perception of the environment certainly took place—in the early years, for example, an inadequate perception of the environment had led to attempts to colonise excessively isolated and physically difficult land. One aspect of this was the changes in ideas about the necessary size of plot required for the *colonia* farmer. At Baraderó´ in 1856, allocations of only 3.2 hectare had been made, at Las Conchas (later Villa Urquiza) in Entre Rios, six-hectare lots were granted in 1853 (Kaerger, 1901). At Esperanza, 20 cuadras, or 33.7 hectares was the standard, while later *colonias* had much larger allocations, as previously mentioned. As in the United States in the period of Homesteading, the idea of what constituted an adequate farm size had to change to meet climatic adversity.

An improvement in the rationality of economic patterns through time, as decision-makers either adjust their practice, or are eliminated, may be presumed to have occurred on the northern edge of settlement, where early *colonias* were established in territory on the fringe of the Chaco, and failed entirely in many cases, while a few remained through moving to cattle-based operations. In the early days, perception of the land possibilities had certainly been defective, because Castellanos, for example, the founder of Esperanza, the first reasonably successful colony, had proposed settlement of the whole north-east of Santa Fe up to 28° Latitude south, well into the southern edge of the Chaco, and inhospitable land for any kind of farming. In the other direction, Gallo (1969) explains the lack of effort in the south of the province as due to faulty information about the land's quality; Burmeister, a leading geologist and Carrasco, a high government official in the province, had given a poor account of the Pampas soil. Brougnes, Lelong, and other visionaries had similarly advocated State participation in settlement in the inhospitable distant parts of the new republic, displaying a general lack of understanding of the agricultural possibilities of the various regions. In 1859, Ramsy Rubert conceived a plan and was granted 200 square leagues by Santa Fe province, for a massive colonisation along the Saldao river, which runs into Santa Fe out of the totally desert province of Santiago del Estero (Hotschewer, 1953) and about this time there were several different favourable reports on the possibilities of settlement and navigation along the river.

A distorted or limited perception was not a prerogative of individuals. The Government of the Argentine Confederation was opposed to the settlement of foreigners on the banks of the San Javier river, major artery of the region, as had been proposed initially by Castellanos, because too great a threat would be posed to national security by foreigners possibly taking control of the river. He was, therefore, forced to use land on the western bank of the Salado, in a difficult and isolated situation. The confederation government also gave excessively small lots— 1858 granting 12 leagues on the Parana, and 24 leagues of land on the Salado, to be settled in lots of only six hectares per family (Schopflocher, 1955).

Eventually all *colonias* were located on reasonably good lands, near transport means and with adequate size plots. But these were not the only changes or even the central ones in the modification of the *colonia* and we must therefore also consider other analyses.

Changing Technological Environment

A second kind of argument focuses on the changes which inevitably occurred over the relatively long period of diffusion of the *colonia*. Changes in the physical environment to which the innovation spread are not so important—the principal direction of spread was to the south and into what should have been easier physical conditions for the establishment of small farms, not the harsher conditions of climate or soil towards the north that would have forced adjustment in farm practices and institutions. Central Santa Fe, the origin, was indeed marginal to the Pampas, but the southern part of the province, like eastern Cordoba and Entre Rios, was good quality land where the apocalyptic sequence of drought, locusts, flooding, and crop diseases that affected the early colonies was not experienced.

Rather than physical changes across space, it is the technical change through time which may be considered to have induced change in the *colonia.* A key element in the changing technology was alfalfa. This crop was grown by the ranchers to feed the introduced pedigree European breeds of cattle, Hereford, Shorthorn and Aberdeen Angus, which were being adopted in the late 19th century. The system adopted by ranchers was one of wheat grown under contracts, of four or five years, followed by alfalfa planted by the tenant farmer with his last wheat, so that on this leaving the alfalfa would be growing up strongly among the wheat stubble : a system that eliminated coarse perennial grasses and weeds by constant cultivation and left the rancher with the bonuses of planted alfalfa and four years of rents. The writers cited so far concur that tenancy was positively associated with the use of alfalfa, and so the deformation of the *colonia* towards tenant farming might be seen as a result of the technical requirements of one crop. Kaerger (1901), saw the relative scarcity of tenant farms in Entre Rios province as a result of the fact that alfalfa does not grow well there. Sternberg (1972) viewed tenancy in southern Santa Fe as promoted by alfalfa needs. But these arguments overlook the powerful spread of tenancy before alfalfa. In the 1870s and 1880s (Table 4.1), tenant farming had become quite important, while alfalfa plantings did not become common in the four-year contracts, until 1897, according to Sternberg himself (1971).

TABLE 4.1: **Percentage of farmers managed by their owners**

Year	BA	Cor	ER	Sta. Fe
1895	49	75	68.6	49.3
1914	30.8	39.5	50	25.6

Source: Sternberg (1971), p. 102, Table 3.

Economic Arguments

Another viewpoint considers not only the technological changes but the economic advantage or disadvantage, profit or loss, in their adoption. We may first examine the economies of an institutional change. If tenancy became a rational change away from sale of land to incoming small farmers, and big landowners sought to hold onto their land, it was because this land was becoming increasingly valuable as a resource, and landowners did not wish to sell it during a phase of rising value. Rising values per hectare of land have been well-documented for the Humid Pampas area in this period, and it only remains to explain why they should have risen. This writer has advanced elsewhere (Morris, 1977) the idea that ranchers were receiving high and improving prices for their products on the London market and this encouraged them to continue in cattle-raising in spite of the competition from the *colonia agricola*, and if we allow some improvement in efficiency of landuse as well, almost certainly a fact since improved stock management was being widely introduced, then there are ample reasons for the rising prices of land that presented a new environment for the *colonia* in the period after 1870, and especially after 1890.

Regional Growth vs. International Epigenesis

Even the economic argument is incomplete, for it is necessary to ask why a cattle-dominated system was presenting so successful a challenge to other systems and notably to the *colonia agricola* within the Pampas in the late 19th century. We may consider the overall process as being, on the one hand an outward expansion of the *colonia agricola*, from Santa Fe, with its production of grains, and on the other an expansion of the worldwide, imperial economy of North-west Europe, and particularly of Britain, which affected the River Plate area generally, and called into being first, the extensive production of hides and tallow in the early period up to 1850, then wool from improved sheep-breeding, and finally after 1880, wheat and improved cattle, exported first as live cattle to Europe, then as meat when refrigeration became possible. In this system of livestock farming, grain production was a useful auxiliary which could be controlled and expanded or contracted as required by the landowners. It has been shown (Morris, 1977) that the centre of cattle improvement innovations was in the *cabanas* of northern Buenos Aires province, both for improvement of sheep and cattle production. In this area, the information about agricultural improvements circulated under the aegis, in part, of the Sociedad Rural Argentina, created in 1865, the leading ranchers' organisation. The *cabanas* brought in the Lincoln sheep, the Hereford, Shorthorn and Aberdeen Angus cattle, but also the ideas on improved hygiene for the animals, controlled breeding, controlled use of pastures and feed supplements, the use of wire fences and windmills (Sbarra, 1973).

The idea that is being suggested here is that of a packing order, a hierarchy, of growth processes. Within this hierarchy, a growth process which was that of a major world system, the British metropolitan system, fed with raw materials by a vast assortment of formal colonies and informal, economic appendages, including the River Plate area, was dominant over the growth process of a much more limited and less powerful institution, the *colonia agricola*. When the two met head on, through the expansion of colonies from Santa Fe and that of modernised beef-producing estancias from Buenos Aires, the dominant concern, the estancia with its links to a world market, determined the fate of the secondary one, the colony. In effect, though the colony came to produce also for world markets and export wheat and maize, it was seen that there was no room for a special agricultural settlement with this sole aim in mind; instead, the small farmer could be accommodated as a subsidiary landuser on the great estancia, and his ploughing of the land for grains be adopted as an element in the rotation of landuse with long grass leys as the dominant use.

This discussion is no longer in terms of diffusion. The expansion of Britain's 19th century economic empire is a process of growth with differentiation, a process of epigenesis[1] whereby the hinterland stretched out further and further (Peet, 1972) with, at the same time, an international division of labour and specialisation of landuse zones. It was the role of regions such as the River Plate, along with Australia and South Africa to be incorporated into this scheme as producers solely of raw materials, hides, tallow, wool, grains, meat. The epigenetic process, organised by a central metropolitan region, dominates the growth process of the *calonia agricola.*

Theoretical Considerations

This leads us on to some thoughts about the current fashion in geographers' studies of growth. When seen in spatial terms, most geographers have subsumed all kinds of growth processes under the general heading of diffusion. Yet it is apparent that at least three levels of growth process are involved in what has been described above, and these may be given an approximate classification as follows :

Level I: Major Systems Growth Processes

Characteristics — Growth occurs with differentiation, an epigenetic process. It may also occur in irregular advances, not by interpenetration of the growth object and its environment, and have a definite frontier.

Examples — British economic empire of 19th century. Formal or informal empires in general. Transnational firms and industries.

Level II: Growth Processes of Institutions and Complex Concepts

Characteristics — Growth without differentiation in an organised manner, though with changing content from one place to another.

Growth may be of diffusion type, i.e. interpenetration of object and environment. Flexibility of content is due to the physical objects of innovation being grouped loosely around a central concept or idea which is abstract.

Examples : Colonia Agricola : Agricultural or industrial revolution, agrarian reform movements, religious reform movements (the Reformation in Christian Religion).

Level III: Growth Processes of Individual Innovations

Characteristics — Growth without differentiation in the usual case, though there may be cases of changing content or use of the innovations;[2] growth may be modelled by diffusion in both time and space dimensions.

Examples: Improved breeds of cattle; wheat; barbed wire; television; cooperatives.

These three levels of growth process are linked, for usually those of Level III are nested within those of Level II, and a similar relation holds between Levels I and II. Spread of the use of Hereford cattle as a Level III innovation was limited by the advance of the new improvements in agriculture, whether in Britain or in Argentina. Beyond these limits, where land was unenclosed, held communally, or did not involve production of crops with animals, there was no need or incentive for introduction of the improved breeds. But the agricultural revolution, which was represented in Argentina by the advance of the new improved cattle breeds, was itself limited by the extent of the commercial and economic empire of Britain. Beyond these limits, it was very difficult for improving movements to take hold, and in the interior of Argentina, for example, as over nearly all of Latin America, they would generally be rejected until modern times.

A final consideration which stems from this view of growth processes is that present day classifications of diffusion processes by geographers have not encompassed a number of important characteristics.

These classification schemes, and the studies on which they are based, lay primary stress on the spatial structure of innovation, i.e. on whether the movement is via a hierarchy or contagion, and on the mechanisms used, whether those of interpersonal contacts, or long-distance information flows such as those of mass media. Specifically, there is no emphasis of the varying *content* of the innovation, and it is this content which determines what kind of model we may apply to it, as well as the manner in which diffusion or growth takes place. It may well be claimed that some studies, for example, those of the spread of cities, of economic modernisation, or of urbanisation, are in fact inappropriate subjects for the use of a diffusion model since they involve the spread, not of a simple or single object, but of a loosely linked set of items which spread more or less together over space and time. Geographers' focus on mechanisms has been wise in the context of search for a precise model of diffusion, but the moment may have come to realise that most growth processes do not conform to the simple model's requirements in terms of unchanging content and spread unconfined by that of related elements in a group.

NOTES

1. By this is meant simply the process of simultaneous growth with differentiation, identified as the form of biological growth processes in the 1830s and replacing the older notion that biological development was by a simple accretion of cells to a pre-formed and complete egg. In geography, some expansion processes, like those of the *colonia agricola*, are essentially growth alone, although complex. Others are epigenetic, involving an increasing distinctiveness of various parts.
2. Interesting cases of changing content are suggested by Isaac (1970) in his study of the diffusion of cultivated crops. Wheat is thought to have been substituted by barley and rye on the N-W fringes of Europe and by Millet and Sorghum towards the hot dry south and west, in a process of spread from a Middle Eastern hearth of domestication.

REFERENCES

Gaignard, R., 1967 : "Origine et Evolution de la Petite Propriete Paysanne dans la Pampa seche Argentine", in *Les Problemes Agraires des Ameriques Latines*, Centre Nationale de la Recherche Scientifique, Paris.

Gallo, E., 1969 : "Occupation de Tierras y Colonisacion agricola en Santa Fe", pp. 92-104, in Alvaro Jara, *Tierras Nuevas*, El Colegio de Mexico, Mexico City.

Hotschewer, C. E., 1953 : *Evolucion de la Agricultura en la Provincia de Santa Fe* Ministerio de Hacienda, Economia e Industria, Santa Fe.

Jefferson, M., 1926 : *Peopling the Argentine Pampa*, American Geographical Society Research Series No. 16, New York.

Kaerger, K., 1901 : *Landwirtschaft und Kolonisation im Spanischen Amerika*, Band 1, (Duncker and Humblot, Leipzig).

Morris, A. S., 1977 : "The Failure of Small-farmer Settlement in Buenos Aires Province, Argentina", *Revista Geografica*, Vol. 85, pp. 63-77.

Peet, R., 1972: "Influences of the British Market on Agriculture and Related Economic Development in Europe Before 1860", *Trans. Inst. of British Geographers*, Vol. 56, pp. 1-20.

Sbarra, N. H., 1973 : *Historia de las Aguadas Y el Molino*, Eudeba, Buenos Aires.

Schopflocher, R., 1955 : *Historia de la Colonisacion Agricola en Argentina*, Editorial Raigal, Buenos Aires.

Semple, K. and Brown, L. A., 1976 : Cones of Resolution in Spatial Diffusion Studies : A Perspective", *Professional Geographer*, Vol. 28, No. 1, pp. 8-16.

Sternberg, R., 1971 : *Farms and Farmers in an Estanciero World*, 1856-1914, Ph. D. Dissertation, Syracuse University.

————, 1972 : "Occupance of the Humid Pampa : 1856-1914", *Revista Geografica*, Vol. 76, pp. 61-102.

White, R. L. and Watts, H. D., 1977: "The Spatial Evolution of an Industry : The Example of Brioler Production", *Transactions, Inst. of British Geographers*, N.S., Vol. 2, No. 2, pp. 175-91.

Wilcken, G., 1873 : *Las Colonias :* Informe sobre et estado actual de las colonias agricolas de la Republica Argentina, Sociedad Anonima, Buenos Aires.

5

ENVIRONMENT AND ADOPTION OF HIGH YIELDING VARIETY OF SEEDS IN MADHYA PRADESH

S. K. SHARMA

THE SOUL of the Green Revolution was the 'miracle seed', often called high yielding variety of seeds. It was proclaimed that yield would increase by four to six folds with the use of these seeds. Even then these seeds could not fascinate the farmers of this state to the extent they did in Punjab. Further, there is wide variation in the use of these seeds within the state itself. Although along with other innovations, adoption of high yielding variety (HYV) of seeds depends on a host of factors, environmental factors play decisive role in this state. Since seed is an essential input, it does not make much difference if one uses HYV or local (*deshi*) seeds in terms of cost, if it is available easily. It involves some technical knowledge. Thus its use is also controlled by socio-economic factors as well as environmental factors. This is evident from the analysis of the extent and spatial concentration of the use of HYV in Madhya Pradesh in perspective of the environmental characteristics. The acceptance of HYV is the result of the interaction between man, his cultural and physical environments.

Among the physical environmental factors which go long way in influencing the use of high yielding variety of seeds climate is at the top. It is because the climate is difficult to be manipulated to the extent the land and soil can be. Soil fertility can be enhanced with the use of fertilisers and land can be levelled. But the climatic conditions cannot be modified to the desired extent. For germination, growth and maturity of each crop a certain combination of temperature, rainfall and moisture regime is essential. Its presence or absence determines the cultivation of that particular crop. To extract the maximum benefit out of the existing climatic conditions, the photosensitive period, strature and the physiological processes should be manipulated in accordance with the existing climatic rhythm. But this rhythm varies widely and seed suitable for one climatic zone may not necessarily be suitable for the other. Problem is more serious in the case of crops grown in hot and wet conditions. Rice provides a very good example which is grown in rainy season in Monsoon Asia. Despite several efforts HYVs of rice could not be as popular as of wheat. Nevertheless, in areas of controlled water supply instead of rain-fed cultivation, HYVs are popular. Contrary to it, HYVs of crops grown in rabi season have received remarkable acceptance. Such environmental problems have been highlighted in the present paper, citing the case of Madhya Pradesh.

Area Under High Yielding Variety of Seeds

The use of high yielding varieties of seeds is comparatively very limited in Madhya Pradesh. Out of the total area under four major grains only one-fifth (20.3%) was sown with the high yielding varieties in 1984-85. It is very low in comparison to the national average of 57.2 per cent. The comparative position of this state with respect to the national averages of the area of cereals under HYV is presented in the Table 5.1.

TABLE 5.1: Area under HYVs of cereals in M.P. and India, 1984-85

(Area in thousand hectares)

Crop	Madhya Pradesh					India
	Area under crop	HYV area as per		HYV area as % of area under crop		HYV area as % of area under crop
		Land-record	Agri. Dept.	Land R.	Agri. Dept.	
Wheat	3597.6	1485.5	1598.2	41.3	44.4	82.9
Rice	4957.3	527.6	1751.1	10.6	35.3	56.9
Jowar	1909.2	198.2	762.4	10.4	39.9	32.4
Maize	849.7	83.1	209.6	9.8	24.7	35.5
Bajra	162.5	31.0	88.0	19.1	54.2	49.3
Total	11476.3	2325.5	4409.3	20.3	38.4	57.2

Sources (i) Commissioner Land Records — Annual Agricultural Statistical Tables of M.P. 1984-85; (ii) Directorate of Agriculture, M.P. - Agricultural Statistics, 1985; and (iii) Tata Services Ltd., Statistical Outline of India, 1986-87.

Thus the state's average of area under HYVs is less than half of the national average (presuming land record data more reliable). Pertinent point to note is that rice is the first ranking crop in terms of area occupied in the state but only one-tenth of it is under HYVs, while national average is well above half of the area. The ratio between HYV area under rice and wheat is 1:3.9. This wide difference in proportionate area under HYV of these two major crops is bound to be manifested through the gross production of these crops. Consequently, despite higher area, production of rice lagged far behind of that of wheat. Total production of rice was 3761 thousand tonnes, while that of wheat 3780 thousand tonnes in 1984-85. Contrary to the general belief, yield rate of rice is (802 kg/ha) much lower than that of wheat (1141 kg/ha) in the state. Another point is of the seasonality in adoption of HYVs. Out of these five crops, only wheat is rabi (winter) crop and others are kharif crops grown in rainy season. Wheat alone contributes nearly two-thirds (63.9%) of the total area under HYVs of these crops. Thus it can safely be inferred that climatic conditions seem to have played significant role in adoption.

Spatial Variation in Use of HYVs

The average situation varies widely. Area under HYVs of rice, wheat and Jowar are presented in Fig. 5.1. Average for these three crops is 21.1 per cent. It ranges from less than 0.1 per cent in Bhairamgarh, Chhindgarh, Antagarh, Keskal (Bastar), Pali, Gohparu and Anuppur (in Shahdol) to 98.7 per cent in Chaura and 97.4 per cent in Kailaras of Morena district. Only 38.8 per cent of total blocks had proportionate area higher than the state average. These blocks are concentrated over: (i) the Chambal Valley, (ii) the Bundelkhand uplands, (iii) part of the Malwa Plateau, (iv) middle Narmada Valley, (v) eastern Satpura, (vi) Tons basin of Rewa, and (vii) the core of the Chhattisgarh basin.

Thus there is seasonal and regional variation in the use of high yielding variety of seeds. Further, it is person specific also. Though there is not much wider variation, yet highest proportion of area under HYVs is on large farms. It has been seen in the case of Madhya Pradesh as a whole (Sharma, 1988). In two separate sample surveys also this notion is confirmed. In first survey 100 farmers lying in seven villages of Sagar district were surveyed with the objective to find out their aptitude towards new technology; while the other survey was conducted in the Bhopal-Vidisha Plateau with the objective to measure the level of agricultural development by Malkhan Singh (1989). Findings of these surveys are presented in the Table 5.2.

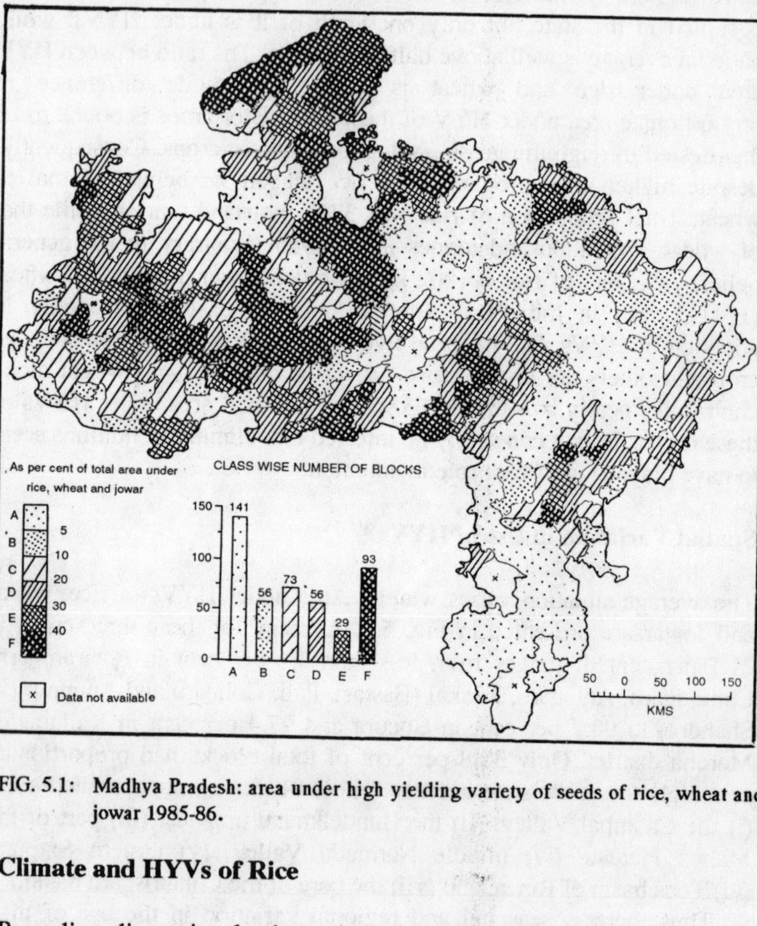

FIG. 5.1: Madhya Pradesh: area under high yielding variety of seeds of rice, wheat and jowar 1985-86.

Climate and HYVs of Rice

Preceding discussion leads to the conclusion that crops grown in dry season have facilitated the use of HYVs, to some extent, while crops, like rice, grown in rainy season, could not be replaced by HYVs. Consequently, very small proportion (only 10.6%) of rice cropped area could be sown with HYVs. Certainly, there are limitations of farmers in obtaining fresh improved seeds every year along with the short-comings of the distribution system (Dasgupta, 1977), which come in the way of popularisation of HYVs of rice in this state also.

Besides them, there are certain environmental constraints which could not be resolved. In the beginning of the 'Green Revolution' rice strains developed by the International Rice Research Institute were tried in the Chhattisgarh basin — the rice-bowl of Madhya Pradesh. They did not succeed. Later on, the Rice Research Centre at Raipur and the Agriculture University of the state developed several other strains claimed to be suitable to the physical conditions of the state (Sharma and Jain, 1987). But all of these HYVs faced common problems.

TABLE 5.2: **Use of high yielding variety of seeds by size of holdings among sample farmers, 1985-87**

Size class	Sagar District			Bhopal-Vidisha Plateau		
	Total Holdings	Holdings using HYV	% age of Total	Total Holdings	Holdings using HYV	% age of Total
Small (below 2 ha)	24	09	35.5	21	13	61.9
Medium (2-4 ha)	21	12	57.1	22	16	72.7
Large (Above 4 ha)	55	38	61.1	100	84	84.0

Most of the HYVs of rice are of dwarf varieties and require long photoperiod characteristics. But the climatic conditions of the eastern M.P. do not match with the requirements of these varieties. Normally, more than 90 per cent of annual rainfall is received in the southwest monsoon season when rice is grown in the state. Rice cultivation is concentrated in southeastern part of the state (Jain, 1988) which happens to be the area of high rainfall (between 100 to 180 cm). In normal years most of the area of this part becomes water-logged and flooded. Farmers have evolved several local varieties suitable to such conditions working under them since time immemorial. Different stages of growth and maturity of plant are adjusted with the existing climatic conditions. Contrary to it, most of HYVs mature in 130-150 days. Therefore, such varieties as IR8 or IR20, sown with the first rains, mature, while the monsoon is active, contrary to the photo-period sensitive traditional varieties which mature as the monsoon retreats. Thus HYVs are not convenient to the farmers as it is difficult to harvest during rains. Further, because of the short straw, the monsoon flood water exceeds the height of the plant, which is injurious to these species. They also require longer photo-synthesis period which is bound to be affected under cloudy monsoon conditions which also adversely affects the yield. The susceptibility of these varieties to pests and diseases under hot and humid monsoonal condition is also one of important considerations. Another characteristics of rainfall of this part is its high variability. Drought years occur after few wet years, with shortage of water even in the growing season. There are not adequate facilities of irrigation to safeguard against insufficient and uncertain rainfall. Research institutes concerned with the development of new varieties of rice are trying to solve such problems as has been discussed by Farmer (1979), yet these new strains could not replace local variety to the remarkable extent.

Along with the HYV, use of fertiliser is recommended. The light sandy soils have comparatively low fertility status and feeding of fertilisers is also necessary in order to balance or to supplement nutrients carried by the floods. There are several problems in the use of fertilisers. First of all, optimum dose of fertilisers, though varies from place to place includes 100 kg of nitrogen, 62 kg of phosphoric acid and 45 kg of potash per hectare of rice under HYV. Majority of the cultivators just cannot afford for it because of tiny landholdings they possess. Consequently, actual use of fertilisers is depressingly low. Average for the Chhattisgarh as a whole was 11.1 kg per hectare in 1982-83 highest being 15.6 kg in Raipur district. This amount is just nominal in view of the recommended dose. It results in comparatively low yields. At the sametime quality of grain is much inferior to several local varieties.

Because of these environmental limitations, HYVs of rice could not be as popular as of the wheat. However, there are 84 blocks where more than one-tenth of the rice area is cropped with the HYVs. Three areas of such blocks are particularly worthmentioning (Fig. 5.2). All of them have developed irrigation facilities. First is the Mahandi-Seonath interflue area of the Chhattisgarh with the canal and tank irrigation facilities. The Wainganga basin and adjoining area of Balaghat is the second one. Both of them are the parts of the major rice-producing tract of the state. Third area of concentration of HYVs is the Chambal basin where rice is insignificant in crop structure and where because of low rainfall irrigation is essential. In fact, rice is an irrigated crop in this area and 100 per cent of HYVs area in Bhind and 98.6 per cent in Gwalior district was irrigated in 1984-85. It means HYVs of rice could be popular in areas where there are facilities of controlled water supply, which are free from hazards of flood and water-logging and have protection against rainfall vagaries. It is because most of the problems enumerated earlier are non-existence there. Application of fertilisers is also competible under such conditions. Consequently, yield rate is also much enhanced. It is very clear from the comparison of situation in two typical areas (Table 5.3). Districts of the Chhattisgarh area are representatives of rice cultivation under uncontrolled rain water, while districts of the Chambal area grow rice under managed water.

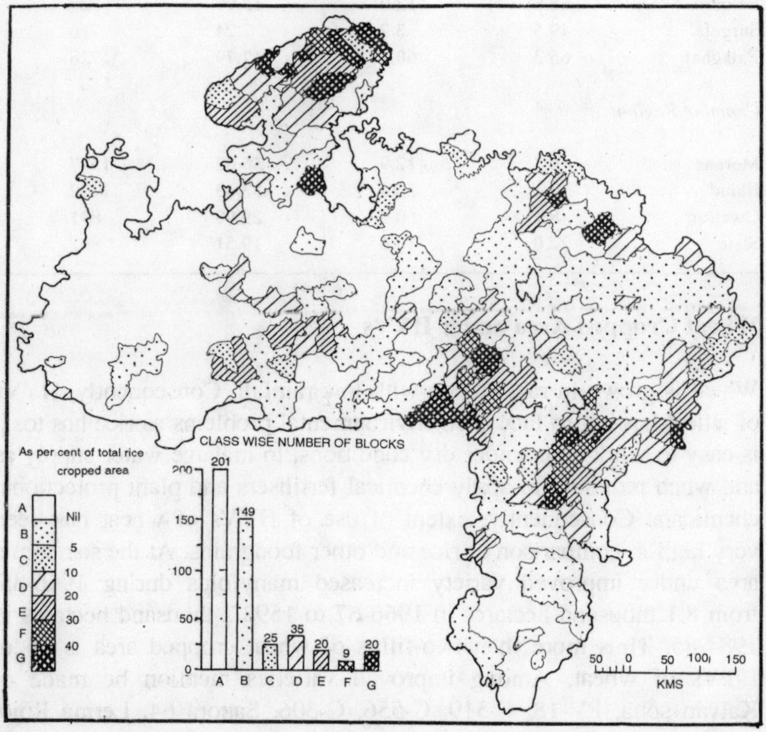

FIG. 5.2: Madhya Pradesh: rice cropped, area under HYV 1985-86.

TABLE 5.3: Use of HYV of rice and fertiliser and yield rates in two typical areas, 1984-85

Region/ District	% of Total cropped area under rice	% of Rice cropped area under HYV	Use of fertilizer (kg/ha of total cropped area)	Yield of rice kg/ha
Chhattisgarh Region:				
Raipur	66.1	25.2	30.17	1217
Bilaspur	66.4	2.1	27.55	880
Durg	47.5	4.2	19.70	1028
Rajnandgaon	43.4	1.8	11.88	828
Raigarh	68.6	12.0	15.99	820
Surguja	49.5	3.2	7.24	716
Balaghat	65.3	60.4	17.79	728
Chambal Region:				
Morena	0.4	12.9	56.68	1577
Bhind	3.1	33.0	33.43	1402
Gwalior	8.1	60.9	29.99	891
State	22.0	10.6	19.51	802

Wheat Cropped Area under HYVs

Wheat is grown in rabi season with low rainfall. Consequently, HYVs of wheat has not to face such environmental problems as rice has to. It is easy to work soils under dry conditions, to manage water supply as and when required, to apply chemical fertilisers and plant protectional chemicals. Consequently, extent of use of HYVs of wheat has been very high in comparison to rice and other foodgrains. At the sametime, area under improved variety increased manyfolds during 1966-85, from 8.1 thousand hectares in 1966-67 to 1598.2 thousand hectares in 1984-85. Thus more than two-fifths of wheat cropped area is under HYVs of wheat. Among improved varieties mention be made of Kalyan sona, PV-18, C-519, C-556, C-306, Sanora 64, Lerma Rojo, S-227, hybrid-65 and RR-21. They give much higher yield than the local varieties (Chourasia and Singh, 1972), though they fetch lower price in the market. Nevertheless, net profit is much higher than the local varieties. HYV of wheat is not only high yielding but also resistant to insects, pests and diseases as well as very fertiliser responsive. With the help of water and fertilisers it can be grown on comparatively low fertile soils. A cumulative result is the high adoptability of HYVs of wheat in the state.

There are 254 development blocks out of 459 with proportion of wheat area under HYVs higher than 40.0 per cent in the state. They form seven extensive belts of high concentration of HYV of wheat (Fig. 5.3). Important among them is the Narmada valley, extending from Narsinghpur in the east to Khargone in the west. HYVs occupy as high as 100 per cent of wheat cropped area in eight blocks of Khargone district and one block (Burhanpur) of Khandwa district. It is noteworthy that wheat is not major crop in this tract except in Hoshangabad. Second area of concentration is the north-western Malwa, spreading over blocks of Mandsaur, Shajapur, Rajgarh, Ujjain, Ratlam, Dewas and Jhabua. Again wheat has not prominent place in crop structure of this area. The Chambal basin comprises the third area of concentration of HYVs of wheat and extends over Morena, Bhind, Gwalior and Datia districts. They constitute part of the major wheat producing region of the state. Use of HYVs is accompanied by the large scale irrigation and high use of fertilisers. Part of the Bundelkhand upland, extending over Chhatarpur, Tikamgarh and Sagar districts is another segment of wheat zone using HYVs to the remarkable extent.

There are three areas of concentration of HYV in the eastern part. One of them extends over the eastern rim of the Chhattisgarh basin from Gariabad block of Raipur to Bagicha block of Raigarh district. Eastern Surguja, just north of it, is its northern extension. Area under wheat is comparatively small in this area, nor there is sufficient irrigation facilities. Nevertheless, usually more than two-thirds of HYV area is irrigated. Even in the least irrigated district of Raigarh more than 91.8 per cent of HYV area is irrigated in 1984-85. The Maikal plateau constitutes another area of HYV of wheat. It covers parts of Balaghat and Mandla districts. Though Balaghat is one of the irrigated districts of the state HYV cropped area of wheat could not be irrigated. Rewa district constitutes third and last area of concentration. More than half of wheat cropped area is under HYV and nearly one-third of it is irrigated.

There is, thus, direct relationship between the proportion of wheat area under HYV and its irrigation. With a few exception, expansion of HYV area has been accompanied with irrigation. It is evident from the fact that more than 71.3 per cent (1059 thousand hectares) of HYV wheat area is irrigated. At the sametime, proportion of wheat area under HYV has been comparatively much higher in non-traditional area of wheat. Thus the physical environment, particularly climate did not present hindrances in its expansion. Rather, farmer is free to water it when it is required. Infact, cultivation of HYV of wheat is basically dependent on irrigation. Proportion of irrigated area of HYV wheat cropped area was above 90 per cent in 21 districts, between 80 and 90 per cent in 3 districts and below 80 per cent in only 6 districts.

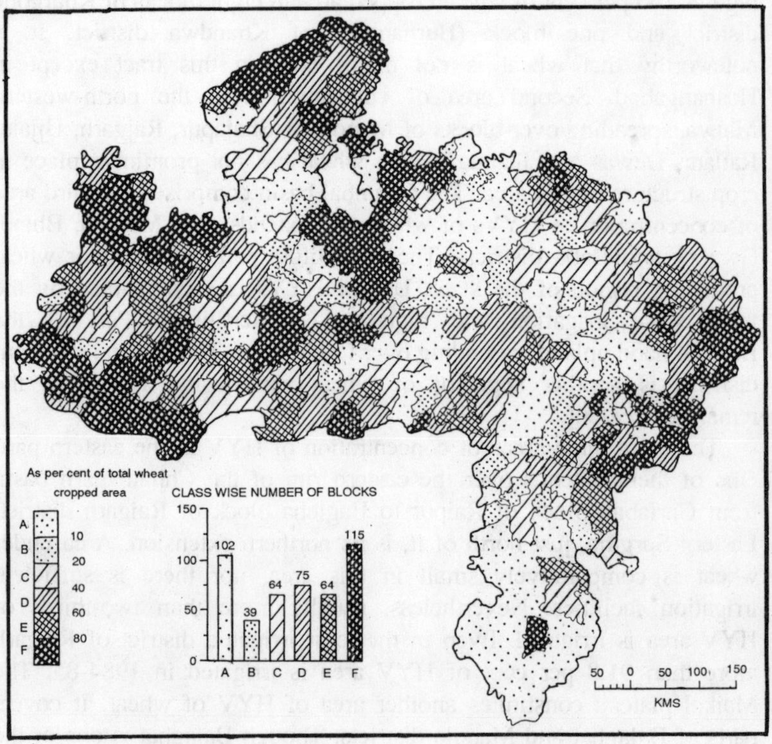

FIG. 5.3: Madhya Pradesh: wheat cropped area under HYV 1985-86.

Controlled water supply has facilitated the use of chemical fertilisers also. For instance, in Tikamgarh per hectare use of chemical fertilisers is 46.24 kgs along with 91.3 per cent of wheat cropped area under HYV. There are ten out of 19 wheat growing districts where per unit use of chemical fertilisers is higher than the state average of 19.51 kgs. Because of this fact, yield and production of wheat increased very rapidly in 'green revolution' period. The state average of yield rate of wheat was only 577 kg per hectare in 1965-66 but reached to 1141 kg per hectare in 1984-85, recording an increase of 97.7 per cent. Remarkable increase occurred in districts where HYV is accompanied with irrigation and fertilisers, which is most often the case. Thus, contrary to rice, adoption of HYV of wheat has also been less problematic because of congenial environmental particularly climatic conditions.

Jowar Cropped Area Under HYV

Jowar is the third ranking foodgrains in the state occupying 8.5 per cent of cropped area in 1984-85. Again, as per the land records only one-tenth (10.4%) of it could be sown with HYVs of jowar. There are only 50 development blocks with higher proportion under HYV than the state average. With a few exception, these blocks are concentrated in the middle Satpura region (Chhindwara, Betul and Khandwa districts) and south-western Malwa region (Jhabua, Dhar, Indore, Ujjain and Dewas districts). These regions form the southern part of the jowar cropping area of the state. The northern Malwa is another half of jowar cropped area but HYV could not be popular there. Infact, jowar is a cheap grain and is grown on comparatively less fertile field which otherwise cannot be sown with superior crops. Because of the cheapness it is neither irrigated nor is provided with fertilisers. As per land records only 0.5 per cent of jowar area under HYV was irrigated in 1984-85. In absence of the application of the package inputs yield rate of jowar could not be enhanced to the extent it was expected. For instance, average yield per hectare was 702 kgs in 1965-68 which rose to only 800 kgs in 1984-85, recording an increase of only 14 per cent. Thus, low yield failed to fascinate the farmers. Further, the phenomenal increase in soyabean area has snatched good land even from jowar. Thus, it is the economics that has come in the way of adoption of HYVs of jowar in the state.

Social Environment and Adoption of HYVs

No doubt, similar to other innovations, HYVs are new to the environment and to the farmers. Their use requires scientific knowledge of the crop management. Therefore, they cannot be used by every one. Only those farmers can use them who have some scientific orientation and aptitude. From this point of view, tribal people stand distinct from the non-tribals. With low literacy and least interaction with outside world tribals lack modern scientific orientation; and thus, they lag far behind in the race of the development including the use of yield raising technology (Sharma, 1989). This state has 12 million tribal people in 1981, being largest in the country. They are concentrated in five zones in the southern, eastern and western parts of the state. The State Government has identified 176 blocks as the Tribal Development Blocks and excepting one all are in these areas. Of the total area under rice, wheat and jowar in these tribal blocks only 7.6 per cent was sown with HYVs as against 21.1 per cent in the state. Out of the total, only 33 (19%) blocks have higher proportion of area under HYV than the state average. Thus very few tribal blocks come forward to use HYVs.

Crop-wise analysis of use of HYVs reveal another fact. Rice is significant crop, occupying nearly one-third of the total cropped area, but only 5.5 per cent of its area is under HYVs in these blocks. The average for the state is one-tenth. Another 5.8 per cent of gross cropped area is under wheat, out of which 38.3 per cent is sown with the high yielding variety of seeds. Thus in comparison of wheat, acceptance of HYV of rice is insignificant. This is true in the case of the entire state also. Jowar occupies only 6.5 per cent of cropped area and only 7.3 per cent of it is under improved seeds. Thus, the level of adoption is considerably lower in tribal areas than the average of the state but the pattern is the same. However, it can be concluded that they find it difficult to adopt agricultural development measures including HYVs.

Conclusion

With the objectives of raising agricultural production measures have been taken to increase the yield. Among such measures, high yielding variety of seeds are of pivotal significance. The use of this along with other yield-raising technology has been influenced by the personal, social and economic characteristics of farmers and by the physical environment of the region. With other things being equal, rainfall goes long way in conditioning the use of HYV. It is evident from the case of rice and wheat. Because of the uncontrollable rhythm and variable nature of rainfall, long rainy season with cloud cover, and recurring phenomena of drought and flood in the rice zone of the state, short strature, early maturing and photo-period sensitive HYVs of rice could not be adopted. Efforts have been made to overcome some of these constraints but remarkable success could not be achieved. Because of low adoption of HYV production of rice could not increase remarkably. However, it is worthmentioning, that in areas where rice is grown under controlled water management remarkable success has been achieved in use of HYVs and fertilisers and also in raising the yield rate. Thus the crucial factor is the management of water supply in adoption of HYVs of rice in this state.

This notion is confirmed by the extent of wheat cropped area under HYVs. Being grown in rainless winter season, wheat is free from the climatic problems and therefore, there is least environmental problem in adopting its HYVs. Requirement of moisture is met by irrigation. Irrigation and use of fertilisers have accompanied the application of HYVs of wheat. In case of jowar economic factors are decisive. Thus, need is to work in two directions. Since, climatic milieu cannot be changed, improved strains suitable to it must be evolved locally rather than importing them from foreign climate. Secondly, attention has to be paid towards water management, which will save the plant from destruction at the times of excessive rains as well as in drought.

REFERENCES

Chourasia, R. R. and Singh, V. N., 1972 : Economics of local and high yielding varieties of paddy and wheat in Panagar village of M.P., *Indian Journal of Agricultural Economics*, XXVII-1, p. 93

Dasgupta, B., 1977 : Indian's green revolution, *Economic and Political Weekly,* XII, No. 6-8, pp. 241-260.

Farmer, B. H., 1979 : The green revolution in South Asian rice fields: Environment and Production, *The Journal Development Studies,* 15-4, pp. 304-319.

Jain, C. K., 1988 : *Patterns of Agricultural Development in Madhya Pradesh,* New Delhi: Northern Book Centre.

Sharma, S. K., 1988 : Adoption of agricultural development measures in relation to size and tenancy of operational holdings in M.P., *Indian Journal of Regional Science,* 20-2, pp. 27-39.

———, 1989 : Use of yield-raising technologies in tribal blocks of Madhya Pradesh, pp. 71-89 in P.D. Tiwari and A.N. Sharma (eds.), : *Tribal Ecosystem and Malnutrition in India*, New Delhi: Northern Book Centre.

Sharma, S. K. and Jain C. K., 1987 : Green revolution and faltering steps of agriculture in the Chhattisgarh region, *Indian Journal of Landscape Systems and Ecological Studies,* 10-1, pp. 41-50.

Sinha, B. N., 1975 : *Modernisation of Indian Agriculture — High Yielding Varieties and Green Revolution,* Bhubaneswar: Eastern Geographical Society.

Singh, Malkhan, 1989 : *Development of Agriculture in the Bhopal-Vidisha Plateau, M.P.,* Unpublished Ph.D. Thesis, University of Saugar, M.P.

6

CORRELATES OF INNOVATIVENESS AMONG FARMERS IN THE GANGA PLAINS

R. RAMACHANDRAN

A WIDE range of farm innovations have been introduced in Indian farming since 1948, when the National Extension Scheme and later the Community Development Programme were started. The three decades that have since elapsed have brought about major changes in the farming methods and practices in all parts of the country. The response of farmers to innovation, however, does vary significantly as between different regions of the country, and even within a single region there are sharp differences as between different farm communities. These differences can ultimately be traced to the level of the individual farmer and his personal characteristics. Despite these differences, the response of Indian farmers to innovation in the farming methods, although slow in the initial decades, has indeed been very substantive in the last decade.

This study explores the differences in the response of individual farmers to innovation in the Ganga Plains. The study is based on field interview of 480 heads of farm households in 12 villages in Meerut district of Uttar Pradesh conducted during September to December, 1975 (Fig. 6.1). The study villages were selected on the basis of two criteria : (a) dominance of a particular social group, and (b) distance from an urban place. Four dominant social groups among farmers in the Ganga Plains selected for the study are : (a) Jats, (b) Rajputs, (c) Gujjars, and (d) Muslims. Three villages were selected with one of the above as a dominant social group. In each of the 12 villages 40 farmers belonging to the dominant social group were interviewed on the basis of a random sample. The villages were also chosen in such a

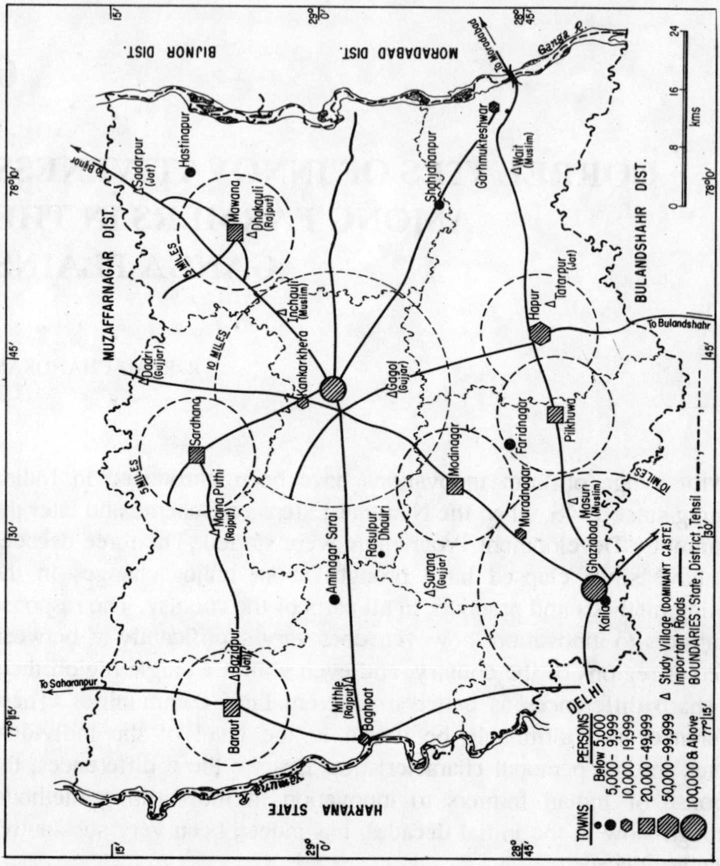

FIG. 6.1: District Meerut—study villages—relative location.

way that half the number of study villages were located within a distance of 10 kilometres from an urban place while the others were located farther away.

Each of the 480 farmers interviewed was asked questions relating to the adoption of a set of 64 farm innovations. Of these 64 innovations 29 were actually adopted by more than 10 per cent of the farmers (Table 6.1).

TABLE 6.1: **List of 29 innovations adopted by more than 10 per cent of 480 farmers interviewed in the study**

Sl. No. Innovations	No. of adopters
Mechanical Innovations:	
1. Tube-well (owned)	269
2. Tube-well water used on hire	202
3. Tractor (owned)	63
4. Tractor used on hire	223
5. Harrow	99
6 Thresher	182
Fertilisers:	
7. Urea	381
8. Ammonium Sulphate	106
9. Ammonium Nitrate	93
10. Diammonium Sulphate	235
11. Calcium	170
12. Superphosphate	66
13. Muriate of Potash	73
14. Mixture of NPK	250
Pesticides:	
15. Gamaxine	348
16. B.H.C.	88
New Crop Varieties:	
Wheat	
17. RR 21	276
18. No. 1553	223
19. No. 308	117
20. No. 1593	88
21. No. 227	84
22. Hira	80
23. No. 553	72
24. Sonalika	67
25. No. 306	62
26. No. K-68	56
Sugarcane	
27. Co 1148	302
New Crops:	
28. Maize	65
29. Potatoes	54

Only one farmer had not adopted even one innovation. On an average, farmers in our study had adopted nine innovations, the maximum number of innovations actually adopted out of 29 innovations being 12. However, over 58 per cent of the farmers had adopted from 10 to 12 innovations each. Farmers adopting less than five innovations accounted for four per cent of the total. These bare facts reveal the extent of change that has occurred over a period of three decades. There is no doubt that agriculture in the study villages has undergone a major transformation and the very basis of traditional agriculture has been altered by the use of new crops, and crop varieties, chemical fertilisers, tube-wells and tractors.

Innovativeness

Sociologists have defined innovativeness generally as the degree to which an individual farmer adopts innovations earlier than others. Innovativeness is a behavioural attribute of farmers and is undimensional, in the sense that those farmers who adopt one set of innovations earlier also adopt most other innovations the same way. If we can measure innovativeness of individual farmers, then it would be possible also to relate this to the farmers' personal characteristics, farm characteristics and exposure to communication channels. Innovativeness is a dependent variable which is in varying measure influenced by independent variables such as the ones mentioned above. An understanding of the relationship between the two is an essential prerequisite to any programme of induced or planned change in Indian farming. This study explores the nature of the relationship between innovativeness and the attributes of farmers and farming characteristics.

Measurement of Innovativeness

As said earlier, 29 innovations covering various aspects of farming have been adopted by at least 10 per cent of the farmers. The aspects of farming covered are : mechanical equipment, fertilisers, pesticides, new varieties of wheat and sugarcane, and new crops. The wide range of innovations studied and the fact that each of them has been adopted by at least 10 per cent of the farmers provide a firm basis for the measurement of innovativeness.

Measurement of innovativeness involves both the number of

innovations adopted by a farmer and the year of its first adoption. The larger the number of innovations adopted, the greater the innovativeness of the farmer. Likewise, the earlier a farmer adopts an innovation, the greater the measure of his innovativeness. The following mathematical experiment is used to measure innovativeness :

$$\text{Innovativeness score} = \sum_{Y_k=1}^{29} (1976 - Y_k)$$

where Y_k is the year of actual adoption of an innovation k, there being a total of 29 such innovations described above. Y_k is assumed to be 1976 when the innovation under consideration has not been adopted by a farmer and the incremental score becomes zero.

The measure of innovativeness, defined above, gives equal weightage to all the 29 innovations, all being considered as equally important *per se*. However, two problems emerge. The first problem has to do with the role of the innovation in the farming economy. Innovations vary in importance in terms of the capital investment involved and their contribution to the gross and net output. The latter aspect is extremely difficult to measure and, in the absence of adequate data, has been ignored. The capital investment involved cannot be directly used as a weighting system, since a variety of other considerations such as economic life of the innovation, its operational costs, etc., have to be taken into account. Unfortunately, for each of data this aspect could also not be taken into account. The second problem in the measurement of innovations arises on account of the built-in weightage given to innovations that are introduced earlier than others. This is so because, in the case of innovations introduced 5 or 10 years earlier, the (1976-Y_i) values would automatically be larger and would count towards the overall innovativeness score. Three considerations are advanced in justification for this built-in weightage is relatively small, (2) those innovations introduced earlier than 1960, are the more capital-intensive innovations such as tractors and tube-wells and added weightage to these innovations would be justified on account of the reasons given in the earlier part of this paragraph; and (3) the adoption of the first innovation by a traditional farmer is a major step forward, which would facilitate the adoption of other innovations later, and hence the built-in weightage is justified.

The innovativeness scores of 480 farmers vary from 0 to 160, the mean value being 67.4. The frequency distribution of the scores shows double peaks—around 45 and 85. This occurs because our sample is not from the same social or locational group. If the samples are drawn from the same community, the frequency distribution would be normal. The fact that the innovativeness scores are not normally distributed may be taken as evidence of significant differences in the adoption behaviours among the social groups and the locational categories. The basic underlying notion here is that communication occurs freely within social groups but not between them and hence the differences in their behavioural patterns.

Social Groups and Innovativeness

The average innovativeness scores vary in a remarkable way as between the four social groups identified in this study. The four social groups fall distinctly into two categories. The first category consists of the Jat and Rajput farmers who have average scores of 79.9 and 74.4. The second category consists of Gujjar and Muslim farmers whose average innovativeness scores are 59.9 and 55.3. The gap between the two groups is very wide while within the categories the gaps are small. The Jats and Rajputs emerge as the most innovative group of farmers while the Gujjars and Muslims are, in a relative score, the least innovative groups. In reality, however, there are many other groups of farmers who may turn out to be even less innovative than the two above. The fact that the social groups emerge as two distinct categories partially explains the double-peakedness in the frequency distribution of the overall innovations scores. However, in the frequency distribution of scores of individual social groups, double-peaks occur in two cases, namely, among Jat and Gujjar farmers; while the frequency distribution of Rajput and Muslim farmers shows a near normal pattern. These differences within social groups may be due to the locational differences among the three village communities involved in each social group and also sociological differences of lower order within each group.

The innovativeness scores show the least dispersion around the mean in the case of Rajput farmers, who form the most homogeneous among the four social groups. Greatest variations occur among the Muslims, followed by Jats and Gujjars (Table 6.2).

TABLE 6.2: Innovativeness and social groups

Social Groups	Innovativeness scores																
	0-9	10-19	20-29	30-39	40-49	50-59	60-69	70-79	80-89	90-99	100-109	110-119	120-129	130-139	140-149	150-159	Total
Jat				3	10	10	21	11	25	21	11	7		1		1	121
Rajput			3	3	11	12	15	30	19	14	8	4	1				120
Gujjar	1	1	6	18	22	15	11	16	18	8	3						119
Muslim	8	3	10	16	18	22	8	10	8	8	4	3	2				120
Total	9	4	19	40	61	59	55	67	70	51	26	14	3	1		1	480

Innovativeness score = $\sum_{1}^{29} (1976-Y_k)$, where Y_k is the year of actual adoption of innovation, there being 29 such innovations.

TABLE 6.3: Innovativeness and location

Location	Innovative Score																
	0-9	10-19	20-29	30-39	40-49	50-59	60-69	70-79	80-89	90-99	100-109	110-119	120-129	130-139	140-149	150-159	Total
Near Urban Places	9	3	13	24	32	31	29	31	28	22	9	7		1		1	240
Away from Urban Places		1	6	16	29	28	26	36	42	29	17	7	3				240
Total	9	4	19	40	61	59	55	67	70	51	26	14	3	1		1	480

Location and Innovativeness

It is generally hypothesised that innovativeness is highest among those groups exposed to urban influences. This hypothesis is not substantiated by the evidence obtained from this study. In fact, innovativeness tends to have an inverse relationship with relative location with respect to urban places. The average innovativeness score for farmers living near urban places is only 63 while the average innovativeness scores for farmers living farther away from urban places is 71.9, the difference being substantial. It is also interesting to note that the highest and lowest innovativeness scores (the tails of the distribution) occur in the first category (Table 6.3). This also brings out the fact that innovativeness tends to vary greatly within the group which lives near an urban centre. This may be partially explained as the result of widening disparities within this zone. On the other hand, the evidence presented here tends to contradict the well-known theory of diffusion of ideas and innovations away from innovation centres, which are usually the urban centres. It is possible to argue, on the basis of the evidence presented in the previous chapter, that in the initial stages of the diffusion process, the urban centres play a key role but, at a later stage, their role is limited. At any rate at the later stages of the diffusion process the roles of adoption would be greater away from the urban centres than near them. These theoretical considerations reflect the complex picture that emerges from Table 6.3.

Personal Characteristics of Farmers and Innovativeness

Rural sociological studies have indicated that innovativeness can be predicted on the basis of an assessment of the personal characteristics of farmers. In order to find out to what extent this hypothesis is relevant to the study villages, six variables were selected. These include: (1) age, (2) education, (3) family type, (4) family size, (5) family members working outside village, and (6) perceived social status of the farmer. Analyses of the data on these variables reveals that while all the variables are related to innovativeness, the nature of the relationship is not always a simple linear relationship, but a complex curvilinear relationship. The younger farmers below 40 years were definitely more innovative (Innovative score 73.8) than the middle-age and older farmers. The middle-aged farmers had an average innovativeness score of 65.6 while the older farmers' score was 64.4. Above the age of 40 innovativeness was not influenced by age.

Farmers' educational attainments were in part related to innovativeness. However, the farmers who had never attended school had nearly the same average innovativeness score as farmers who had completed primary level education (65.2 and 65.5) respectively. Farmers who have completed matric-level education or have gone in for further education had a substantially higher innovativeness score (78.4). Clearly education is relevant in the context of adoption of innovations, but the educational level appears to be not primary education but secondary level education (Table 6.4).

The majority of the farm families had a joint family set-up, while the nuclear families were few, and extended families even fewer. The extended families had the lowest innovativeness score (51.4), while the nuclear families came close to the average (65.8). The joint families were more innovative than the other two types (70.0). This probably reflects the level of resources available to the family, the joint families being better off in this respect. Analysis of the data on family size tends to support this view. The innovativeness score tends to increase with family size up to a point, namely, 14 members, beyond that it tends to decline. The family group with 10-14 members had the highest innovativeness score of 71.9 while families having 7-10 members had the lowest score of 65.8.

An important feature of farm families in Meerut is the phenomenon of family members working outside the village. There were three categories of families having members working outside the village. The first category consists of families, one or two members of which are working in other villages in or outside Meerut district, presumably as farm workers. These families had lower than average innovativeness score (63.0). The second category includes families whose members are working either in Meerut city or in other urban (nearby) centres in the district. These families tend to have very high innovativeness scores (78.5). The third category consisting of families whose members are working in centres outside Meerut district had low scores (64.6). Presumably, in the third category the contact between the family and members working in the city was not important enough to influence the farm and its working.

TABLE 6.4: Innovativeness and educational level

Educational Level	Innovative Score																
	0-9	10-19	20-29	30-39	40-49	50-59	60-69	70-79	80-89	90-99	100-109	110-119	120-129	130-139	140-149	150-159	Total
Never attended school	7	3	9	18	25	21	18	18	25	20	12	4	1				181
Primary incomplete	1		2	6	6	11	8	17	14	8	4	2	1				80
Primary complete	1	2	7	12	13	10	9	4	5	3	1						67
Matric incomplete		1	4	8	13	9	9	16	14	14	1	3	1				93
Matric complete			2		2	2	7	4	7	2	3	2					31
Diploma holder					3	1	2	1	2	1		1					11
Degree holder				1		2	1	2	4	1	3	1		1		1	17
Total	9	6	24	45	62	56	54	62	71	49	24	13	3	1		1	480

TABLE 6.5: Innovativeness score and size of holdings

Size of holding in bighas	Innovativeness Score 0-9	10-19	20-29	30-39	40-49	50-59	60-69	70-79	80-89	90-99	100-109	110-119	120-129	130-139	140-149	150-159	Total
0-9	—	—	—	—	3	—	1	4	4	2	1	2	—	—	—	—	17
10-19	1	2	3	1	1	4	6	4	2	3	2	2	2	—	—	—	33
20-29	2	—	6	13	16	11	8	10	17	7	5	2	—	—	—	—	97
30-39	1	1	1	10	7	11	16	12	5	12	1	1	—	1	—	—	79
40-49	1	1	3	4	11	7	5	6	10	4	3	1	—	—	—	—	56
50-59	2	—	2	5	6	7	7	6	9	9	8	—	—	—	—	—	61
60-69	1	—	—	1	5	6	3	9	9	6	3	2	—	—	—	1	46
70-79	—	—	3	2	4	4	2	4	2	2	1	2	—	—	—	—	26
80-89	1	—	—	3	1	3	1	3	5	—	1	2	—	—	—	—	20
90-99	—	—	—	—	1	1	—	3	—	1	—	—	1	—	—	—	7
100+	—	—	1	1	6	5	6	6	7	5	1	—	—	—	—	—	38
Total	9	4	19	40	61	59	55	67	70	51	26	14	3	1	—	1	480

The perception of status by the interviewers was clearly related to innovativeness. The low-status farmers had an average score of only 50.7, while the high status farmers had an average score of 77.6. It is reasonable to presume that status and adoption of some least of the innovations are directly inter-related. For example, the farmers owning tractors or tube-wells also attain higher status by virtue of this fact. It is true that in some cases, the acquisition or maintenance of status becomes an important consideration in the adoption of innovations.

Farm Characteristics and Innovativeness

The characteristics of farms, such as size, intensity of cropping and degree of commercialisation of farming have a definite bearing on the adoption of farm innovations. The status of farms in terms of tenure or whether farming is done with the aid of irrigation and some are important variables, but are not relevant to the farming situation in the study villages, as all the 480 farms are owner-cultivated and all farms are irrigated either from canals or tube-wells. Nevertheless, there are significant differences in farm size and intensity of farming which merit detailed analysis. Tables 6.9 to 6.11 shows the relationship between innovativeness and selected farm characteristics such as : (1) size of holding, (2) intensity of cropping, (3) degree of commercialisation. The latter two variables are represented by surrogates, namely, sown area as portion of net area of holding and per cent of total cropped area in cereal crops. In all cases there are significant but complex patterns of relationship.

The smallest farmers having less than 10 bighas (two acres) are the most innovative group, their average innovativeness score being 80.3. However, innovativeness declines with increasing farm size until the farm size approaches 50 bighas (10 acres). Beyond this the farmers become again very innovative (score of 76.0 for the category of 60-69 bighas). The higher level of innovativeness persists beyond this level of farm size, despite one exceptional category (Table 6.5). The noise in these tables results from differences in family type and family size.

The intensity of farming is related to innovativeness. However, the relationship here is reciprocal. Farmers with low intensity of farming also have low innovativeness scores (54.7). Where the farming intensity is just above average, innovativeness is highest (85.8). Thereafter, with increasing intensity of farming, innovativeness declines, but nevertheless, the level of innovativeness continues to be above the overall average of 67. However, one major conclusion that emerges from the analysis is that the most innovativeness farms are not necessarily the most intensely cultivated farms.

The degree of commercialisation is directly related to innovativeness. Farms where more than 70 per cent of the cropped area is devoted to cereal crops have very low innovativeness scores (45.9). Innovativeness is highest where 60 per cent of the cropped area is devoted to non-cereal crops (79.8). Most of the major agricultural innovations are related to cereal crops, particularly wheat. Yet, their adoption depends, to a large extent, on the ability of the farmer to finance the investment involved in tractors and tube-wells. This comes mainly from the cash crops, particularly sugarcane, maize, rice, potatoes and so on. This underlies one of the many basic contradictions in our farming systems.

Mass Media Exposure and Innovativeness

Farmers in the study villages are exposed to a variety of mass media, such as television, radio, newspaper, cinema and so on. However, not all farmers were exposed to each of these media nor was the degree of exposure the same. In this section the degree of exposure to various mass media and innovativeness of farmers has been examined.

Exposure to Television and Innovativeness

The frequency with which farmers saw television programmes in general, and the frequency with which the "Krishi Darshan" programme on TV was seen, are both related to innovativeness. Farmers who had seen television programmes frequently and farmers owning television sets had very high innovativeness score (over 80), while farmers who had never seen television programmes had very low scores 64.9. A similar pattern obtains with regard to the exposure to "Krishi Darshan" programme, although in this case, farmers viewing it frequently had a slightly lower score than farmers who had seen it rarely. Incidentally, among those who saw the "Krishi Darshan" programme, some happened to be the least innovative and exposure had no impact on these individuals. They, however, are an exception.

Exposure to Radio and Innovativeness

Innovativeness is directly related to exposure to radio programmes in general and to the *Krishi Gosthi* programme in particular. The farmers who listened to the *Krishi Gosthi* programme frequently had very high average innovativeness (79.5), while those farmers who never heard this programme had very low scores (54.9). Farmers owning radio sets and listening to radio sets owned by others had high scores as well (72). Farmers who never listened to any radio programme expectedly had very low scores (54.6). While it would be unrealistic to assume that the *Krishi Gosthi* programme provided all the information that the farmers needed for the adoption of new methods of farming, it would, however, be reasonable to assume that this programme does provide the basic information for creating awareness of modern methods of farming. The evidence presented here does clearly show the influence of radio on farmers (Table 6.6).

Exposure to Cinema and Documentary Films and Innovativeness

A majority of the farmers in the study villages had seen a commercial cinema at least once and these farmers had a higher innovativeness score (68.7), in the case of farmers who had seen cinema rarely and 75.2, in the case of farmers who saw movies frequently. Those farmers not exposed to cinema had very low scores (64.4). The influence of cinema is evident (Table 6.7). On the other hand, documentary films on agriculture and related problems have a direct bearing on adoption of farm innovations. Those who claimed they saw documentaries frequently had innovativeness scores of 83.3, while those who have rarely seen it had an average score of 79.1. Farmers not exposed to documentary films had below average scores (64.9). It is interesting to note that fewer farmers were exposed to documentary films than the commercial cinema, yet the influence of the former is more pronounced than the latter.

Exposure to Daily Newspapers and Magazines and Innovativeness

Farmers reading daily newspapers frequently had an incredible average innovativeness score of 101.7. There were 47 such farmers. Newspaper reading is related to education and even economic status. The combined influence of all these variables is indeed well demonstrated by the usually high average scores. The farmers not exposed to newspapers had scores of less than 65.9, although there were a few farmers even within this category who were highly innovative. The

TABLE 6.6: Innovativeness and exposure to Krishi Gosthi programme on radio

Exposure to Krishi Gosthi	Innovative Score																
	0-9	10-19	20-29	30-39	40-49	50-59	60-69	70-79	80-89	90-99	100-109	110-119	120-129	130-139	140-149	150-159	Total
Never heard	8	3	14	33	42	29	19	18	11	12	4	5	1	1			200
Rarely	1	1	4	4	14	23	24	30	31	26	13	8					179
Frequently			1	3	5	7	12	19	28	13	9	1	2			1	101
Total	9	4	19	40	61	59	55	67	70	51	26	14	3	1		1	480

TABLE 6.7: Innovativeness and exposure to cinema

Exposure to Cinema	Innovative Score																
	0-9	10-19	20-29	30-39	40-49	50-59	60-69	70-79	80-89	90-99	100-109	110-119	120-129	130-139	140-149	150-159	Total
Never seen	7	2	13	17	30	26	22	29	28	23	10	7					214
Rarely	1	2	6	22	27	28	27	27	35	23	11	7	3	1			220
Frequently	1			1	4	5	6	11	7	5	5					1	46
Total	9	4	19	40	61	59	55	67	70	51	26	14	3	1		1	480

pattern of influence of this media is closely measured in the exposure to weekly newspapers wherein again one finds higher innovativeness associated with greater exposure to weekly magazines. The number of farmers exposed to daily newspapers and weekly magazines form less than 30 per cent of the total.

Summary

Based on the number of innovations adopted by individual farmers and also the year of their first adoption, an innovativeness index has been defined. Innovativeness of farmers as measured by this yardstick has been correlated to: (a) social groups, (b) location, (c) farmers' personal characteristics, (d) farm characteristics, and (e) mass media exposure.

The average innovativeness scores vary in a remarkable way as between the four social groups identified in this study. They fall distinctly into two groups—the first group consisting of Jat and Rajput farmers having very high average scores and the second group consisting of Gujjar and Muslim farmers having low innovativeness scores. In the real work, however, there are many other possible groups of farmers some of whom may be even less innovative than the groups discussed here.

It is generally hypothesised that innovativeness is highest among farmers exposed to urban influences. This hypothesis is not supported by the evidence obtained from field interview. In fact, farmers living in villages farther away from urban centres were generally more innovative than farmers living near urban places. However, near urban places, there is great variation among farmers in innovativeness, indicating widening disparities in this zone.

Farmers' personal and family characteristics such as age, education, family type, family size, family members working outside of village and perceived social status of farmers were found to be related to innovativeness. However, the nature of the relationship is not always simple linear relationship but a complex curvilinear relationship. The characteristics of farmers, such as size, intensity of cropping and degree of commercialisation have a definite relationship with the degree of innovativeness of farmers. These relationships again reveal not simple but complex patterns.

7

SOCIO-DEMOGRAPHIC DETERMINANTS OF DIFFUSION OF AGRICULTURAL INNOVATIONS*

NOOR MOHAMMAD

ALTHOUGH RAJ (1977) maintains that with the growth rate of 2.5 per cent in a country where population has also been growing of the same rate there is a virtual stagnation of economy, and Azad and Garg (1977) opine that foodgrains increase has been much more than increasing population during the years 1951-71, the fact remains that during the last two-three years the agricultural production in India has increased to a large extent and that the country has not only become self-sufficient in supply of foodgrains but has some buffer-stock as well. This growth in agricultural production is, by and large, attributed to the technological change and spatial diffusion of agricultural innovations, which have taken place during the last one and a half decades. But neither the rate of agricultural growth nor the diffusion of agricultural innovations is uniform all over the country. It is due to varied environmental, social, psychological, demographic and economic conditions in the country which affect the process of technological change and diffusion of agricultural innovations. Hence, there is a pressing need to identify and measure the impact of these factors on the adoption of new farm practices. Such studies will help the planners in understanding, controlling and ultimately tackling the problems more efficiently in planning the introduction of innovations.

In order to ascertain the trend of diffusion of agricultural

* This paper is a part of the report of the University Grants Commission's Project on "Technological Change and Spatial Diffusion of Agricultural Innovations in Trans-Ghaghara Plain" undertaken by the Author.

innovations along with the impact of aforesaid factors on it, various sociological (Sachchidanand, 1972; Rajgopalan and Singh, 1971; Van de Ban, 1960; Bose, 1962), economic (Chaudhary, 1979; Singh, 1967; Basu 1978), psychological (Gross, 1942; Pemberton, 1936; Marsh and Coleman, 1966), and geographical studies (Hagertrand, 1958; Bowden, 1965; Rogers, 1962) have been done in the Western countries. Recently several empirical studies have also been conducted in our own country (Mohammad, 1974; Ramachandran, 1975; 1979; Misra, 1968) but no exhaustive study has been done in this connection in the Saryupar Plain. Hence, the author has chosen this region to study the problem.

The Study Area

The Saryupar Plain with an area of 33,227 sq. km. extends from 26°N to 28°24′N Latitude and 81°3′E to 85°E Longitude. It includes the districts of Bahraich, Gonda, Basti, Gorakhpur and Deoria of north-eastern Uttar Pradesh (Fig. 7.1). It has a homogeneous relief with a general slope from north and north-west towards south and south-east. The region may broadly be divided into three physiographic units, viz. the Tarai, the Khadar and the Central Plains. The tarai zone having a gentle slope towards the south and south-east extends all along the northern boundary of the region in the foothills of the Himalayas. The khadar region extends all along the rivers Ghaghara and Rapti and could be divided into Ghaghara Khadar and Rapti Khadar. The former is less fertile as compared to the latter. The central plain is broadly divided into upper central plain and lower central plain.

The region enjoys a sub-humid climate with three well-marked seasons, i.e. summer, rainy and winter. The average annual rainfall varies from 110 cm to 150 cm and more than half of it is received in only two months (July and August) of the rainy season (Fig. 7.1). The amount of rainfall decreases from south-east to north-west. Very little amount of rainfall is received during the winter months but it is very beneficial for the Rabi crops if it is free from hailstorms. The summer season is almost dry as very insignificant amount of rainfall is received during this season. However, the rainfall is slightly variable and the coefficient of variability of rainfall, increasing from south-east to north-west, varies from 16 to 22 per cent (Mohammad, 1978).

Hence, successful agriculture considerably depends upon irrigation. The main sources of irrigation are wells, ponds and tube-wells.

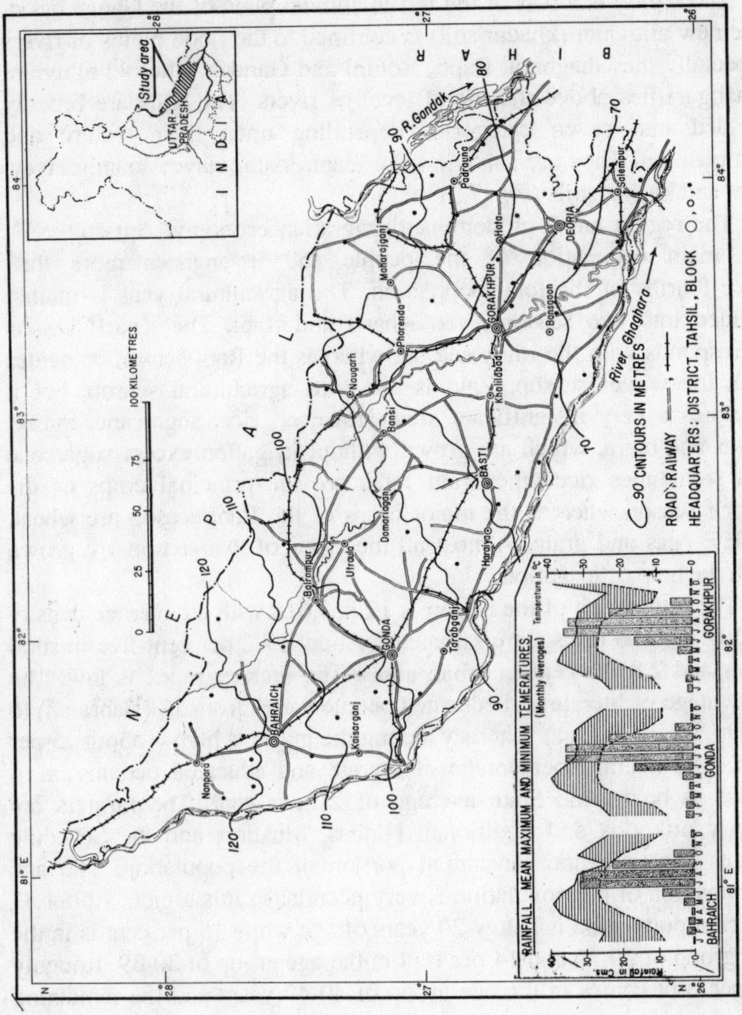

FIG. 7.1: Saryupar plain : geographical setting.

The region is a part of the fertile alluvial plain of the Ganga basin. The new alluvium (khadar soil) is confined to the flood plains of rivers especially the Ghaghara, Rapti, Rohini and Gandak. The old alluvium (bhangar) lies above this flood level of rivers. The soils are broadly divided into seven categories depending upon their texture and composition. They are sandy, sandy loam, loam, clayey loam, clayey, bhat and kachar soils (Fig. 7.2).

The region has a predominantly agrarian economy. Agriculture is the main occupation of the people and it engages more than three-fourths of the total population. The agricultural year is mainly divided into two seasons, viz. Kharif and Rabi. The Kharif season corresponds with the rainy season, whereas the Rabi season coincides with the winter season. Zaid is the third agricultural season, but it occupies a very insignificant area. However, rice, sugarcane, maize, jowar and bajra, which are grown without irrigation except sugarcane and sometimes rice when rain fails, are the principal crops of the Kharif season whereas the major crops of the Rabi season are wheat, barley, peas and grain. Almost all the crops of this season are grown with the help of irrigation.

The population of the region is 12,863,618 with an average density of 387 persons per square kilometre. About 97.2 per cent live in rural areas and 2.8 per cent in urban areas. The literacy index is low. The percentage of literate and educated people varies from 12 (Bahraich) to about 20 (Gorakhpur). Literacy among the males is high – about 25 per cent. The average percentage of literate and educated persons is 16 which is below the State average of 21 per cent. The farmers are mostly orthodox and traditional. Hindus, Muslims and the Schedule Castes constitute the substantial portion of the population. The age composition of the population is very peculiar in this region. About 47 per cent population is below 20 years of age while 15 per cent is in the age group of 20-29 and 14 per cent in the age group of 30-39. Roughly 10 per cent comes in the age group of 40-49 years and the remaining 14 per cent is in the 50 and above age group.

Sources of Data

It is a predominantly rural region with 18,896 inhabited and 1824 uninhabited villages. It is true that the sampled study cannot be as good as the study of the entire population but keeping the time, scope and funds in view it was practically impossible to collect primary data

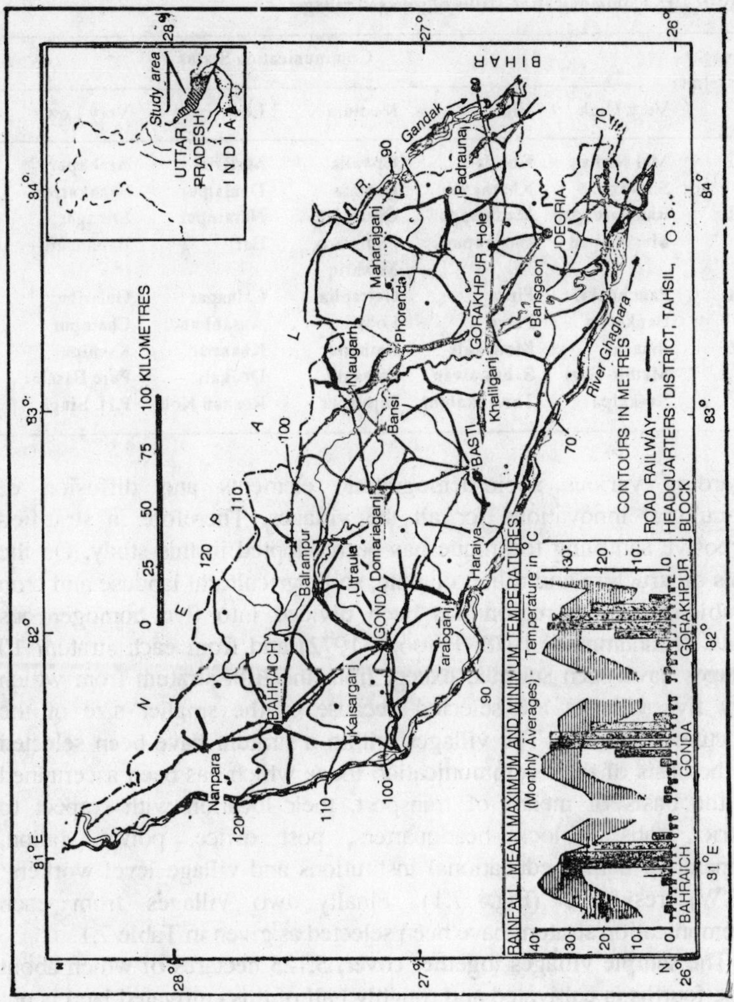

FIG. 7.2: Saryupar plain : homogeneous regions.

TABLE 7.1: **Communication status of selected villages**

Homogen-eous Strata	Communication Status				
	Very High	High	Medium	Low	Very Low
I	Mal Malia	Khoria	Urwalia	Materia	Arakhpur
	Siswari	Kharhatia	Bhainsa	Daulatpur	Shankarpur
II	Rahmatnagar	Madhwapur	Qiyamjato	Narainpur	Srinagar
	Mustafabad	Bhajanpur	Phulwaria Mashriq	Bari	Babhnauli
III	Panchpakri	Pipra	Gujraulia	Lalnagar	Gulariha
	Bankatwa	Parsa	Lohati	Ausankundi	Chainpur
IV	Jhansa	Modhauli	Ratnapur	Khanpur	Kashipur
	Mathia Gazi	Sahsarawan	Dubauli	Deokali	Pure Basalat
V	Sankalpa	Tawakkalpur	Bijwapur	Roshan Kola	P.H. Singh

regarding various socio-demographic elements and diffusion of agricultural innovations for all the villages. Therefore, a stratified purposive sampling technique has been adopted in this study. On the basis of structure and relief, climate, soil, agricultural landuse and crop combinations, the region has been divided into five homogeneous strata. (Mohammad, 1978; Farooqi, 1972) and from each stratum 10 villages have been selected except from the fifth stratum from which only five villages are selected because of the smaller size of the structure (Fig. 7.2). The villages within a stratum have been selected on the basis of their communication index which has been ascertained on the basis of means of transport, their location with respect to district, tahsil, block headquarters, post office, police station, cooperative banks, educational institutions and village level workers' (VLW) residence (Fig. 7.1). Finally two villages from each communication stratum have been selected as given in Table 7.1.

The sample villages together cover, 5,125 hectares of which about three-fourths is cultivated and roughly half of this cultivated land is put to double cropping. The irrigation facility on the whole is inadequate and only 50 per cent of the cultivated land is irrigated. The total population of the villages is 20,138 of which about 87 per cent is engaged in agricultural activities and the remaining are in service, and caste occupation. The educational facilities are inadequate and only 14 per cent of the village population is literate and educated.

Literacy among the males is much higher than among the females. Hindus and Muslims are the two major religious communities but the caste structure is very complex. People belonging to more than a dozen castes are living in these villages. The age structure of the village population is almost similar to the age composition of the population of the region. The farmers are, by and large, traditional and orthodox.

During the last 15 years several agricultural innovations have been introduced in the region but only four of them have been chosen for the present study. They are chemical fertilisers, improved agricultural implements, plant protection chemicals and new varieties of seeds. Although environmental, economic, social, psychological and demographic factors affect the technological change and diffusion of agricultural innovations, only socio-demographic factors have been taken in this paper.

There are 3,748 households living in 3,155 occupied residential houses. Since this size of households is very large, it was also not possible to collect primary data for all the households. Hence, one-third households (1,249) were selected on a random basis. The field work was done in 1974-75 during both the agricultural seasons and various types of relevant data were collected at four levels, viz., district, block, village and farmer. The district and block level data were collected from the respective headquarters, while the data at village and farmers level were collected through personal interviews of articulate farmers who were the functional heads of the families and were instrumental in decision-making in agricultural matters.

Methodology

The entries of the survey schedule of each respondent were thoroughly checked in order to see whether there is any omission of relevant facts and also inconsistency in the answers. The mistakes were removed before processing the data. Different score points were given to various socio-demographic factors and the degree of adoption of selected agricultural innovations. Finally, standing score of each respondent was worked out with respect to each socio-demographic element as well as adoption of innovation. With the help of individual adoption score average adoption index of each group of farmers in different socio-demographic factors was calculated. However, the data were arranged in frequency tables and variables were categorised. These tables consist of those used for statistical analysis and others which helped in proving or disproving the hypotheses.

For the analysis of data, a statistical scheme is prepared. Analysis of variance and "t" tests are used in order to test various hypotheses.

Socio-Demographic Factors v. Adoption of Agricultural Innovations

> If we know what a society's culture is, including its particular systems of values and attitudes, we may predict with a fairly high degree of probability whether the bulk of its members will welcome or resist a particular innovation.
>
> —*Ralph Linton*

The importance of a social system among the elements of diffusion, viz. innovation, communication channel, social system, spatial system and time, hardly needs any emphasis. The adoption of an innovation involves decision-making on the part of farmers. This decision takes into account not only the economic advantages accruing from adoption but also depends upon social setting in which he is placed. Hence, the behaviour, action and characteristics are very much influenced by socio-psycho-cultural factors which include among others religion, caste, mores, law, public opinion, arts and education. The social phenomenon is a function of various cultural forces. Different sociological and socio-psychological studies (Sorokin, 1928) give more importance to these factors than others like geographical, biological and economic ones. In diffusion studies also (Bose, 1966; Singh, 1965) it has been interpreted that socio-psychological correlates are more influential in the context of technological change and diffusion of agricultural innovations. However, in the present study an attempt has been made to find out the relationship between diffusion of agricultural innovations and education, training, caste, age, value-orientation, family size and occupational structure of the people.

Education v. Adoption of Agricultural Innovations

The role of education in the process of adoption of innovation is obvious. It is one of the chief instruments through which a society socialises its members and brings desirable changes in the social life of its people. Education is, in fact, the aggregate of all the processes by means of which a person develops abilities, attitudes and other forms of behaviour of positive values in the society in which he lives or it is a social process by which people are subjected to the influence of selected and controlled environment so that they may attain social competence and optimum individual development (Bear, 1947). And as such it could be argued that education influences the course of behaviour of both society and its individuals.

Since the adoption behaviour of farmers is a learning process involving knowledge, attitudes and skill regarding agricultural innovations, it cannot be exception to it. Several studies conducted in India and abroad (Chaudhary, 1979; Mohammad, 1976; Rahudkar, 1962) regarding the technological change and diffusion of innovations have proved the close relationship between the rate of adoption of educational status of the respondents. It is, therefore, hypothesised in this case also that :

> The rate of technological change and diffusion of agricultural innovations is positively related with the educational status of the farmers.

In this study the broad concept of education including formal and informal sources of knowledge, skill and attitudes has been used in order to assess the educational status of the farmers. Scores were given to various standards of education. Finally the scores of individual farmer were added up so as to give his educational status. The scores of the respondents vary from 0 to 44. On the basis of their individual scores the farmers were classed into the following four groups :

Group A	High education	> 20 scores
Group B	Medium education	20.10 scores
Group C	Poor education	< 10 scores
Group D	No education	0 score

According to the above grouping only 13 per cent of the total respondents belong to high education group and roughly one-fifth to medium education group. The largest share goes to poor education group. However, about two-thirds of the respondents have either poor education or are illiterate.

The average adoption index of each education group is worked out with the help of individual adoption scores of the respondents of the group. In order to test the differences in the average adoption of different educational groups, the analysis of variance technique is used and the emerging results are given in Table 7.2.

TABLE 7.2: Analysis of variance of educational status

Sources	Degree of freedom	Sum of squares	Mean sum of square	'F'
Between Classes	3	1009.2652	336.4218	78.9463
Within Classes	1245	5318.2272	4.2614	
Total	1248	6327.4924		

The calculated value of 'F' is 78.9463 which is much higher than the value given in 'F' test table and hence, the test is significant at one per cent level.

Table 7.3 shows the average adoption index of each educational status group. It is clear that the average adoption index of the high education group is 3.8965 which is much higher than the average adoption index of illiterate farmers. Therefore, it is a good indicator of relationship between education and adoption of new techniques. However, the average adoption index of medium and poor education groups are 2.1873 and 0.9887 respectively (Fig. 7.3). Thus it may be concluded that with the increase or decrease in educational status, the rate of adoption of new practices also increases or decreases. "T" tests indicate that this difference was not just by chance and it is significant at one per cent level.

TABLE 7.3: Education v. average adoption index

Sl. No.	Group	No. of Farmers	Average Adoption Index	Groups Compared	Difference in Average Adoption Index	S.D.E.	
1.	High Education	163	3.8965	1-2	1.7092	0.1232	13.8725
2.	Medium Education	267	2.1873	2-3	1.1985	0.1417	8.4572
3.	Poor Education	485	0.9887	1-3	2.9078	0.6964	4.1755
4.	Illiterate	334	0.5439	1-4	3.3526	0.9824	3.4125
				2-4	1.6434	0.2522	6.5167
				3-4	0.4448	0.0758	5.8712

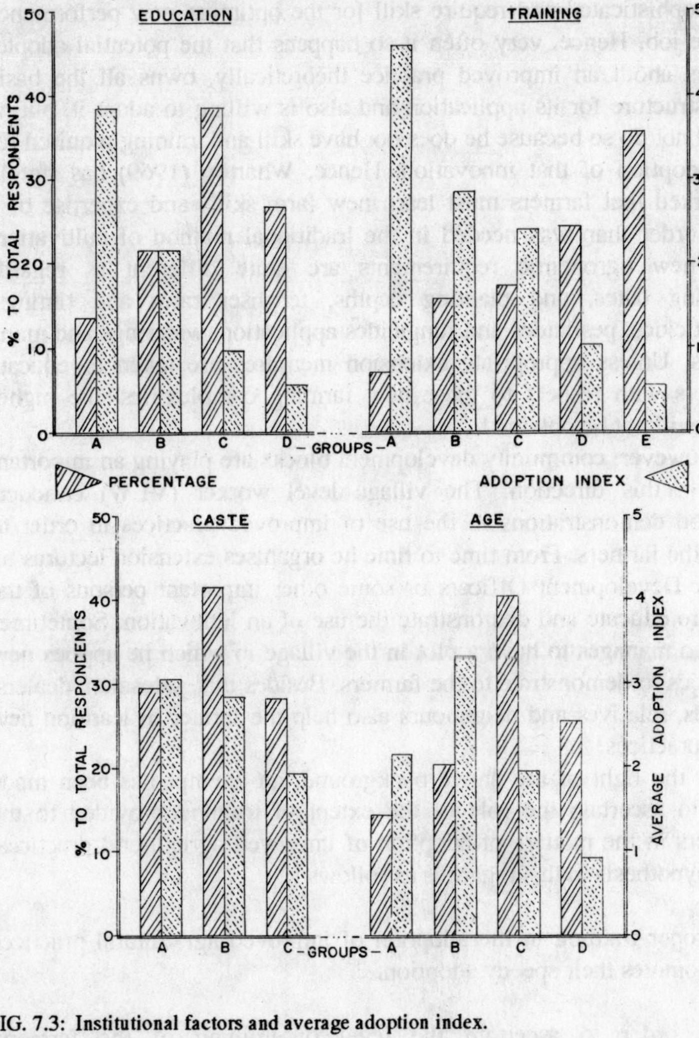

FIG. 7.3: Institutional factors and average adoption index.

Training v. Adoption of Agricultural Innovations

The importance of training in the process of technological change and adoption of agricultural innovation hardly needs any emphasis because it is more specific and practical in nature involving learning by doing.

There are several agricultural innovations which are very complicated and sophisticated and require skill for the optimum new performance of the job. Hence, very often it so happens that the potential adopter knows about an improved practice theoretically, owns all the basic infrastructure for its application and also is willing to adopt it, but he could not do so because he does not have skill and training required for the adoption of that innovation. Hence, Wharton (1969) has rightly remarked that farmers must learn new farm skills and expertise of a high order than was needed in the traditional method of cultivation. The new agronomic requirements are quite different as regards planting dates, and planting depths, fertiliser rates and timings, insecticides, pesticides and fungicides application, waterings and many others. Unless appropriate extension measures are taken to educate farmers with respect to these new farming complexities, the higher yield will not be obtained.

However, community development blocks are playing an important role in this direction. The village level worker (VLW) conducts method demonstrations in the use of improved practices in order to train the farmers. From time to time he organises extension lectures by Block Development Officers or some other important persons of the field to educate and demonstrate the use of an innovation. Sometimes he also manages to have a plot in the village in which he applies new methods to demonstrate to the farmers. Besides this, salesmen, dealers, friends, relatives and neighbours also help the farmer in learning new farm practices.

In the light of the above background, an attempt has been made here to ascertain the role of the extent of training provided to the farmers in the matters of adoption of improved agricultural practices. The hypothesis in this regard is as follows :

> Proper training in the adoption of improved agricultural practices promotes their speedy adoption.

In order to ascertain the level of training of the farmers, information regarding their formal and informal technical education and training, their frequency participation in agricultural courses, seminars and group discussions organised by various extension agencies in the village, or block headquarters or outside the block were collected. In addition to this, sources of information through which he had learnt about the improved agricultural practices adopted by him were also ascertained. The comparative scores were given to all these items and finally the total score of all the respondents according to their performances and responses was worked out. The adoption scores of the respondents vary from zero to seven, but the number of farmers with more than six scores was very small. However, the individual scores were arranged in descending order and divided into five groups in order of priority as given below :

Group A	High group in training	> 4.5 scores
Group B	Medium group in training	4.5-3.0 scores
Group C	Poor group in training	2.9-1.5 scores
Group D	Very poor group in training	1.5 scores
Group E	No training	0 score

In order to assess the standing of different groups of farmers, the average adoption index of each group was calculated and compared with one another. The analysis of variance is used and the results are given in Table 7.4. According to 'F' test table, an 'F' value of 4.61 is sufficient for significance but the calculated value is 127.4875 which is much greater than table value. Hence, the groups differ in their average adoption index according to their training standards. The difference is significant at one per cent level.

TABLE 7.4: **Analysis of variance of training**

Sources	Degree of Freedom	Sum of Square	Mean Sum of Square	'F'
Between Classes	4	2001.9106	500.4177	127.4875
Within Classes	1244	4883.5708	3.9257	
Total	1248	6885.4814		

To understand more clearly the pattern of adoption of different training groups and their differences in average adoption index "t" tests were employed and the results are given in Table 7.5 which shows that there is a positive correlationship between the training standards and adoption of new farm practices. The respondents belonging to the high group in training have the highest adoption score and vice versa. However, the training standard of the respondents is very poor. There are only 87 farmers (6.96%) who have adequate training and as a result their average adoption index is also high (4.5738). Roughly one-third of the total farmers have no training at all and their adoption index is the lowest (0.5291). About 60 per cent of the total respondents have poor or very poor training standard and their average adoption indices are 2.4113 and 1.0217 respectively. There is little difference in the adoption index of medium and poor groups (Fig. 7.3). However, it is obvious from Table 7.5 that all individual differences were significant at one per cent level and with increase in training standard, adoption score also increases and vice versa. And this proves our hypothesis.

TABLE 7.5: **Training v. average adoption index**

Sl.No.	Group	No. of Farmers	Average Adoption Index	Groups Compared	Difference in Average Adoption Index	S.D.E.	't'
1.	High Group (A)	87	4.5738	1-2	1.7094	0.5901	2.8967
2.	Medium Group (B)	195	2.8644	2-3	0.4531	0.0886	5.1162
3.	Poor Group (C)	214	2.4113	1-3	2.1625	0.2124	10.1817
4.	Very Poor Group (D)	309	1.0217	1-4	3.5521	0.2076	17.1122
5.	No Training (E)	444	0.5291	1-5	4.0447	0.2219	18.2266
				2-4	1.8427	0.2370	7.7766
				2-5	2.3353	0.2531	9.2256
				3-4	1.3896	0.3360	4.1358
				3-5	1.8822	0.2277	8.2657
				4-5	0.4923	0.1300	3.7872

Caste v. Adoption of Agricultural Innovations

The Indian social system is inconceivable without caste which influences all other aspects of life. Caste, joint family and village community are the basic social institutions of the Indian society. It is one of the dominant themes in any discussion of economic development and growth (Sorokin, 1928). It plays a major role in technological change and adoption of agricultural innovations. The value-orientation and behaviour pattern of the village people are very much influenced by the ways of life of the respective caste groups. The general notion is that people belonging to higher castes are more progressive and modernised in their values and behaviour than middle and low-caste people of society. As a result they are more adoptive. Hence, the following hypothesis is framed in this connection.

The rate of adoption of agricultural innovation is higher among high-caste people as compared to other castes.

In order to test this hypothesis, the caste composition of the sample farmers was worked out on the basis of census of India criteria and the respondents are classified into three groups, i.e. high caste, middle caste and low caste. With the help of the individual adoption score of respondents the average adoption index of each caste group is calculated and is given in Table 7.7. For statistical testing analysis of variance is used and the results are given in Table 7.6. The "F" value is 57.3458 which is much higher than this value given in "F" test table and hence, the test proved to be very significant.

TABLE 7.6: **Analysis of variance of caste**

Sources	Degree of Freedom	Sum of Square	Mean Sum of Square	'F'
Between Classes	2	530.7698	265.3849	57.3458
Within Classes	1446	6691.7988	4.6278	
Total	1448	7222.5686		

'T' test is used in order to understand more clearly the pattern of adoption of different castes and their difference in average adoption index. The results are given in Table 7.7 which reveals that 't' tests are significant at one per cent level in all the cases.

TABLE 7.7: **Caste v. average adoption index**

Sl. No.	Groups	No. of Farmers	Average Adoption Index	Groups Compared	Difference in Average Adoption Index	S.D.E.	't'
1.	High Caste	374	3.1574	1-2	0.2792	0.1618	11.7253
2.	Medium Caste	521	2.8782	2-3	0.9548	0.1175	8.1227
3.	Poor Caste	354	1.9234	3-1	1.2340	0.1236	9.9872

It may further be observed from Table 7.7 that there is a slight difference in the number of high-caste people, and low-caste people but there is a considerable gap between the average adoption index of these groups. The average adoption index of high-caste farmers is 3.1574 while that of low-caste farmers is 1.9234. The average adoption index of middle-caste farmers who account for roughly two-fifths of the total respondents is 2.8782 (Fig. 7.3). The results clearly reveal that adoption of innovation is positively related with the caste structure of society, i.e. higher the caste, the higher would be the adoption rate.

Age v. Adoption of Agricultural Innovations

In the process of adoption of new ideas and practices, the age composition of the population plays a very important role and diffusion of agricultural innovation is no exception to it. The presumption is that at a comparatively younger age people are more receptive to new ideas and practices (Mohammad, 1977). At an advanced stage people find it difficult to change from age-old practices, and they resist adoption of innovation (Sachchidanand, 1972). For the assessment of association between the age group and rate of adoption the sample farmers are put into following four groups:

Group A	Young age farmers	< 30 years
Group B	Lower middle age farmers	31-40 years
Group C	Upper middle age farmers	42-50 years
Group D	Old farmers	> 50 years.

About one-fourth of the farmers belong to the old group. The percentage of young farmers who are instrumental in taking a decision regarding adoption of innovation is only 14, whereas the upper middle age group constitutes about two-fifths of the total respondents. The lower middle age group accounts about 20 per cent of the sample farmers and as such the middle age group accounts for 60 per cent of the total respondents. However, with the help of the individual adoption score of the farmers, the average adoption index of each group is calculated. To test the hypothesis proper, average adoption indexes of the groups were compared by employing analysis of variance techniques. Table 7.8 shows the results of this analysis which explains that the calculated "F" value is much higher than the value of 'F' in 'F' test table. It proves that the groups differ in their adoption index to a very significant extent.

TABLE 7.8: Analysis of variance of age

Sources	Degree of Freedom	Sum of Square	Mean Sum of Square	'F'
Between Classes	3	224.0021	74.6674	34.7824
Within Classes	1245	2672.6415	2.1467	
Total	1248	2896.6436		

In order to understand more clearly the pattern of adoption of different age groups and the difference in their average adoption index 't' is used which is significant at one per cent level in all the cases. Table 7.9 shows that the average adoption index of lower middle aged farmers is highest (3.3112) whereas it is lowest among the old farmers (0.8786). The average adoption index of young farmers is slightly lower (2.1473) than lower middle aged farmers, because the farmers of this age group are too young to take the risk of adopting an innovation (Fig. 7.3). However, the above analysis proves that age composition, by and large, is inversely related with the adoption of innovations, i.e. with increase in age, the rate of adoption decreases and vice versa.

TABLE 7.9: Age v. average adoption index

Sl. No.	Groups	No. of Farmers	Average Adoption Index	Groups Compared	Difference in Average Adoption Index	S.D.E.	't'
1.	Group A	176	2.1473	1-2	1.1639	0.0857	13.5879
2.	Group B	254	3.3112	2-3	1.3664	0.1151	11.8726
3.	Group C	501	1.9448	1-3	0.2025	0.0428	4.7311
4.	Group D	318	0.8786	1-4	1.2687	0.3993	3.1767
				2-4	2.4326	0.3404	7.1472
				3-4	1.0662	0.1727	6.1737

Value-Orientation v. Adoption of Agricultural Innovations

The significance of social values and norms to the socialisation process and inter-relationship of the people hardly needs any emphasis as it has been well recognised by social scientists (Davis, 1956). The values refer simply to the accepted or required behaviour of a person in a particular situation (Davis, 1959). In other words it refers to the rules of social game observed by the actions while playing their social roles in a social system. They are the behavioural guidelines to the individuals and groups in the network of social relationship. The adoption behaviour of an individual, group and society is no exception to it. If we know the culture of a society together with its systems of values and attitudes, we could very easily predict with a very high degree of certainty regarding the acceptance or rejection of an innovation by its members (Linton, 1962).

With this view it has been decided to find out the correlation between the social values of cultivators and their adoption of agricultural innovations. On the basis of values of the cultivators they have been grouped into two categories : (a) progressive, and (b) traditional. These words have been used in a comparative sense. The progressive stands for more innovative, more modern, more developed or more economically rational farmers towards men and matters, whereas the traditional stands for the reverse conditions. Hence, the progressive farmer is one who views innovations more favourably and adopts new ideas more rapidly than the farmer with traditional values. The former welcomes innovations and the latter resists them. However, four important value dimensions, viz., success-security, secular-sacred, individualism-familism and past-future were taken in the present study with a view that a progressive farmer has success, secular, individualism and future orientation in his outlook and values, and the traditional farmer cherishes security, sacred, familism and past orientation values. Several diffusion studies have established the importance of these values to the practice adoption by the farmer (Rogers, 1962; Dube, 1958). The hypothesis built up in this connection is as follows :

> The rate of adoption among progressive farmers is higher than traditional ones.

In other words the farmers of success, secular, individualism and future orientation tend to adopt agricultural innovations more early and rapidly than the farmers of security, sacred, familism and post-value orientation. In order to assess the value-orientation of the farmers, value-loaded statements on items of interests to the farmers were framed. On the basis of the response of the farmers, their total scores were worked out in respect of all the four dimensions. The individual scores vary from 5 to 23. These scores were arranged in descending order in the case of each dimension and the respondents were divided into two categories, i.e. traditional having lower scores and progressive having higher scores.

On the basis of the individual adoption score of each respondent in each category the average adoption index was worked out and is given in Table 7.10 which also shows the differences in average adoption index between progressive and traditional farmers with regard to all the dimensions by making use of "t" tests. Table 7.10 further shows that the number of progressive farmers is high in first and third value dimensions and is less in other two. In the fourth dimension the progressive farmers constitute only 20 per cent of the total farmers, while in the first they account for 59 per cent of the respondents. However, on the whole the progressive farmers account for 44 per cent of the total respondents and 56 per cent are traditional farmers. Table 7.10 further indicates that the average adoption index of the progressive farmers is higher in all the value-dimensions, and the differences between the average adoption index of the two groups are 1.7168, 1.1229, 2.1881 and 1.1893 respectively (Fig. 7.4). The difference in average adoption index is significant at one per cent level in all the four dimensions which proves our hypotheses.

TABLE 7.10: Value-orientation v. average adoption index

Sl. No.	Groups	No. of Farmers	Average Adoption Index	Difference in Average Adoption Index	S.D.E.	't'
1.	Success	736	2.8922	1.7168	0.2407	7.1316
	Security	513	1.1754			
2.	Secular	438	3.0156	1.1229	0.1013	9.8721
	Sacred	811	1.8927			
3.	Future	793	3.4235	2.1881	0.1888	11.5921
	Past	456	1.2354			
4.	Individualism	253	1.9875	1.1893	0.1466	8.1134
	Familism	996	0.7982			

Occupation v. Adoption of Agricultural Innovations

Occupation is another determinant which may affect the rate of adoption of agricultural innovations. Although the predominant majority of the respondents in this study was of farmers, which had to be so as the main focus of the study was to understand the social dimension of agricultural innovations, there were some farmers who were engaged in some subsidiary occupations, as also some of them continued to practise their traditional occupation, although it had ceased to be their mainstay, such as the Brahmins – some of whom carries on their priestly functions – potters, Harijans, weavers, tailors, barbers, carpenters, etc., and some of them were engaged in petty

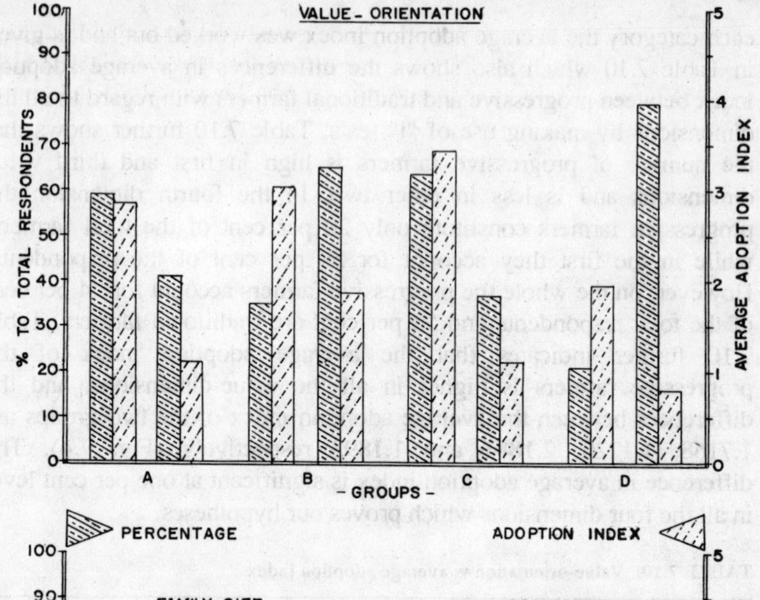

FIG. 7.4: **Institutional factors and average adoption index.**

business, whereas there were some whose other members of the family were in service. However, on this basis of their subsidiary occupations the respondents have been classified in the following groups:

Group A	Farming and service
Group B	Exclusively farming
Group C	Farming and business
Group D	Farming and caste occupation.

The general observation is that among the four kinds of farmers differentiated in this way, those who combine farming with service show the highest adoption index. Therefore, in the present study, an attempt has been made to understand the relationship between type of occupation and adoption of agricultural innovations. The average adoption index of each group has been worked out with the help of individual adoption score of farmers, the average adoption of these groups is studied and a comparison between them made by using the analysis of variance technique. The results are given in Table 7.11 which shows that the calculated value of "F" is 6.2183 which is more than the table value. Hence, test is significant. In order to understand the difference clearly "t" test is carried out by grouping the different groups into pairs. The results are given in Table 7.12 which shows that it is significant at one per cent level in all cases except in cases at 1-2 and 3-4.

TABLE 7.11: **Analysis of variance of occupation**

Sources	Degree of Freedom	Sum of Square	Mean Sum of Square	'F'
Between Classes	3	20.9253	6.9751	6.2183
Within Classes	1245	1396.5165	1.1217	
Total	1248			

Table 7.12 further shows that group A consists of only 125 farmers (10%) and their average adoption index is highest (3.1475). About 69 per cent of these respondents are exclusively farmers, their average adoption index is 2.9876 which is slightly less than that of group A. The adoption index of group A is more because the members of a family who are in service come in contact with more and more people and are more mobile and experienced. They also assist their families in getting agricultural innovations in time. The farmers of groups C and D who together account for one-fifth of the total respondents have lower average adoption indices which are 1.1236 and 0.9876 respectively (Fig. 7.4).

Family Size v. Adoption of Agricultural Innovations

Old agrarian societies of the Orient are characterised as familistic and the whole social organisation and the values and beliefs of the people are centred around the social institutions. Most of the families especially in the countryside are large in size and extended or joint in composition. Parents generally live with their married sons and their families. By and large, it is the family and not the individual which is the unit of social action. The thought and action of the individual members in the family in a large measure are influenced by the family, as a single entity. The personality of a member of a large family is somewhat different from that of a member coming from a small family. Generally a large family finds it more difficult to arrive at a decision regarding the adoption of an innovation in comparison with a small or medium-size family.

TABLE 7.12: Occupational structure v. average adoption index

Sl. No.	Group	No. of Farmers	Average Adoption Index	Groups Compared	Difference in Average Adoption Index	S.D.E.	't'
1.	Group A	124	3.1475	1-2	0.1549	0.1764	0.9064**
2.	Group B	861	2.9876	2-3	1.8640	0.1001	18.6213*
3.	Group C	101	1.1236	1-3-	2.0239	0.4112	4.9099*
4.	Group D	163	0.9876	1-4	2.1599	0.3172	6.3805*
				2-4	2.0000	0.1511	13.2362*
		1249		3-4	0.1360	0.1122	1.2127**

* Significant at one per cent level.
** Significant at five per cent level.

With this view an attempt has been made in the present study to find out the relationship between the size of family and adoption of agricultural innovations. On the basis of size of family, households were arranged in descending order and divided into the following groups :

Group A	Small size	< 5 members
Group B	Medium size	5-10 members
Group C	Big size	11-15 members
Group D	Very big size	> 15 members

With the help of the individual adoption score, the average adoption index of each group is worked out. To test the hypothesis, the average adoption indexes of groups were compared. This is done by employing the analysis of variance technique. The results are given in Table 7.13. The calculated "F" value is 5.1327 which is slightly more than the table which proves that the groups of farmers showed differences in adoption which could not be accounted for by random errors.

TABLE 7.13: **Analysis of variance of family size**

Sources	Degree of Freedom	Sum of Squares	Mean Sum of Squares	'F'
Between Classes	3	15.1394	5.0464	5.1327
Within Classes	1245	1224.0840		
Total	1248			

In order to understand more clearly, the pattern of adoption of different family size groups and their difference in average adoption index, "t" test is employed and the results are give in Table 7.14. "t" test is significant at one per cent level in all the cases except in the case of 2-3 and 1-3. Table 7.14 further shows that the largest number of sample farmers belong to medium-size family. They account for roughly two-fifths of the total respondents and their average adoption score is the highest (3.1258). About 34 per cent (425) respondents have big families and their average adoption index is lower than the former group. The average adoption index is lowest (0.8693) among the respondents having very big families (Fig. 7.4). The main reason, perhaps, is that responsibilities are divided and it is comparatively difficult to arrive at a common decision with regard to adoption of new ideas and practices. In the case of small-size families they do not have enough workforce to change over to a new practice. However, the above analysis proves that there is a positive relationship between the size of a family and adoption of agricultural innovations. The small and very big families are less adoptive of agricultural innovations as compared to medium and big size families.

Conclusion and Suggestions

There is no doubt that the present study is too brief to make generalisations with regard to the impact of socio-demographic factors on technological change and spatial diffusion of agricultural innovations at global or national level, but the study will, perhaps, be able to point out some of its salient features.

TABLE 7.14: **Family size v. average adoption index**

Sl. No.	Group	No. of Farmers	Average Adoption Index	Groups Compared	Difference in Average Adoption Index	S.D.E.	't'
1.	Group A	225	1.0513	1-2	2.0745	0.6509	3.1872
2.	Group B	487	3.1258	2-3	0.1382	0.1242	1.1126
3.	Group C	425	2.9876	1-3	1.9363	2.2602	0.8567
4.	Group D	112	0.8693	1-4	0.1820	0.0916	1.9876
				2-4	2.2565	0.2948	7.6523
				3-4	2.1183	0.7326	2.8913

The socio-demographic factors play a very vital role in adoption of agricultural innovations. All the seven selected socio-demographic factors, viz. education, training, caste, age, value-orientation, occupation and family size, show a significant and positive relationship with extent of practice adoption except that of age which has shown an inverse relationship. Hence, it is suggested that those steps which attempt to improve knowledge, skill and attitude of the farmers with regard to adoption of innovations should be given serious attention by the change agents, research workers, educators and development administrators. Social factors should be given more importance than economic factors as the former appear to be relatively more influential. The following steps should be taken in this connection.

Education is the core of all the socio-cultural factors. In fact, the process of diffusion of new ideas and practices involves five stages, viz. knowledge, interest, evaluation, trial and adoption and the entire process signifies the acquiring of knowledge, skill and attitude on the part of adopters. Therefore, education should not be spread but made compulsory up to higher secondary for all in the country. The illiterate farmers must be provided education facilities particularly the facility of agricultural education. Social and adult education should be revived properly and strengthened. The educated people should be attracted towards farming. The agricultural graduates and post-graduates should be given incentives to take up the profession of farming. Finally, the mass media of education like film shows, television and radio programmes, visits to the farms by experts, demonstrations and kisan tours should be activated and made more meaningful and attractive.

Training particularly in the field of farming plays a very vital role in adoption of agricultural innovations. The trained farmer shows a high degree of innovation and adoption than the untrained one. Hence, it is suggested that a sound and reality-related policy of imparting training to farmers with regard to various improved agricultural practices should be formulated. The educational institutions especially the local ones should be involved in this scheme of training farmers. The VLWs, extension officers and high officials having training in this respect should be invited from time to time to give training to farmers.

Social values and norms also play a significant role in the adoption of agricultural innovations. Hence, attempts should be made through various measures to bring about change in the social values and norms, although it is difficult and different, and it will take considerable time. The decision to adopt or reject a new farm practice in a traditional society is greatly influenced by important social institutions like caste, village community and joint family. The low-caste people are less adoptive as compared to high-caste people. Hence low-caste people should by given various kinds of incentives to adopt agricultural innovatives. Besides this, the inferiority complex which prevails among them must be removed by a spirit of social equality and justice. Age composition also affects the adoption of agricultural innovations. The young farmers are more innovative and adoptive. Therefore, measures should be initiated to give leadership to the young people also or they should be given due importance in decision-making in farming. Further, family size and occupational structure also play a significant role in the process of adoption of new farm practices. It has been observed that a farming group in service is more innovative. Hence, as far as possible local jobs should be provided to some members of the families making up such groups. Likewise very big and small families whose adoption index is low should also be encouraged.

REFERENCES

Ashukant, 1968: "Farm credit, Danger of Downgrading Co-operative Role," *The Madras Journal of Co-operative*, Madras, p. 25.

Barnea, J., 1953-54: "Economic Implications of Electrification in under development countries," *Economic Development and Cultural Change*, Vol. 2, pp. 371-79.

Basu, A., 1978: *Technology Possibilities of Indian Agriculture*, Calcutta, p. 41.

Bell, C. and Zusman, 1976: "A Bargaining Theoretic Approach to Crop Sharing Correlates", *American Economic Review 66*, 4, pp. 78-87.

Bose, S. P., 1961: "Characteristic of Farmers who adopt Agricultural Practices in Indian Villages, *Rural Sociology*, 26, pp. 138-145.

———, 1965: Peasant Values and Innovation in India, *American Journal of Sociology*, Vol. 67, pp. 552-60.

Bowden, L. W., 1965: *Diffusion of Decision to Irrigate*, Department of Geography, Research Paper No. 97, University of Chicago.

Chaudhary, K. M. and Maharaja, M., 1966: "Acceptance of Improved Practices and their Diffusion Among Wheat Growers in Pali District of Rajasthan", *Indian Journal of Agricultural Economics*, 21, pp. 163-64.

Clive, B., 1977: "Alternative Theories of share cropping. Some test using Evidence from north India", *Journal of Development Studies*, 13, pp. 317-26.

Dasgupta, S., 1963: Innovations and Innovators in Indian Village, *Man in India*, 1043.

Desai, D. K., 1966: Technological Change and its Diffusion in Agriculture", *Indian Journal of Agricultural Economics*, No. 1, XXI.

Emery, F. E. and Oeser, O. A., 1958: *Information Decision And Action: Psychological Determinants of Farming Technique*, Melbourne.

Flieglgel, F. C. and Klvin J. E., 1966: *Attributes of Sociology*, Vol. 72, p. 244.

Freeman, C. and Saxena, R. M. (Ed.), 1961: "Economic Status and Adoption of New Agricultural and Home Practices: A study of Three Villages" in *Sociology, Social Research and Social Problems in India Bombay.*

Marsh P. C. and Coleman L. A., 1966: "The Relation of Farm Characteristics", *Rural Sociology* No. 3-4, Vol. 20, pp. 240-91.

Mohammad N., 1974: "The Diffusion of an Agricultural Innovations in Haryana", *The Geography*, No. 1, Vol. XXI, Aligarh.

———, 1976: "Technological Change and Diffusion of Agricultural Innovations, *The Geographer*, No. 1, Vol. XXIII, Aligarh, p. 4-14.

———, 1978: "Impact of Economic Factors on Diffusion of Agricultural Innovations in Trans-Ghaghra Plain, *Geographical Review of India*, Calcutta.

———, 1978: *Agricultural Landuse in India*, Delhi.

Mohammad, N., and Majeed, A., 1979: "Impact of Socio-Economic Factors and Diffusion of Agricultural Innovations", in Mohammad, A. (Ed), *Dynamic of Agricultural Development*, Delhi, pp. 151-74.

Pansi, V. G. and Singh, D., 1966: "Promotion and Assessment of Technological Change in Indian Agriculture", *Indian Journal of Agricultural Economic*, Vol. XXI.

Poffenbergor, A. T., 1961: *Principles of Applied Psychology*, Bombay, p. 398.

Rahudkar, W. B., 1962: "Farmers Characteristics Associated with the Adoption and Diffusion of Improved Farm Practices", *Indian Journal of Agricultural Economics*, Vol. XVII, pp. 82-85.

Rangaswamy, P. et al., 1972: *India's Changing Farmers*, Allahabad. Ray, G.F. and Ulharmann, 1979: *The Innovation Process in the Energy Industries*, London.

Roy, P. et al., 1968: *Agricultural Innovations Among Indian Farmers,* Hyderabad.

Savale, R. S., 1966: "Technological Change in Agricultural Development: Role of Irrigation and Cropping Pattern in Agricultural Development", *Indian Journal of Agricultural Economics*, Vol. XXI, pp. 199-208.

Shetty, M. S., 1966: "Inter-Farm Rates of Technological Diffusion in Indian Agriculture", *Indian Journal of Agricultural Economics*, Vol. XXI, pp. 189-98.

Sorokin, P., 1928: *Contemporary Sociological Theories*, New York, p. 514.

Wharton, C. R., 1969: "The Green Revolution Comucopia or Pandora's Box", *Foreign Affairs*, No. 3, Vol. 47, p. 3.

Wilkening, E. A., 1950: A Socio-Psychological Approach to the study of Acceptance of Innovations Farming, *Rural Sociology*, pp. 352-64.

8

ADOPTION OF MODERN FARM TECHNOLOGY IN MADHYA PRADESH

C. K. JAIN

THE PERIOD of the annual plans, 1966-67 to 1968-69, witnessed the formulation of a new strategy for agricultural production that was favoured by three circumstances. First, the implementation of the "Package Programme" showed that farmers were inclined to adopting new technology, provided that it was demonstrated and the benefits properly explained. Second, new high yielding exotic strains of wheat and paddy, which responded well under irrigated conditions, and the application of heavy doses of fertilisers were introduced and adopted in India after extensive field trials. Third, the Government of India adopted the policy of assured remunerative prices, so that farmer's efforts to increase production would not be inhibited by the fear of undue depression in prices. The new strategy consisted of introducing high yielding varieties of cereals, accepting the policy of irrigation for optimum production, increasing the availability and use of scientific inputs, arranging access to credit and assuring remunerative prices (Sharma, 1982).

The development of high-yielding crop varieties, the improved methods of crop fertilisation, use of pesticides, improved iron ploughing implements, threshers, tubewells, tractors irrigation pumps, etc. constitute the inputs of new farm technology. These techniques are yield increasing and help in enhancing the farm productivity, if they are supplemented by introduction of adequate water supply, mechanical techniques and improved agrarian structure (Singh 1976, Dubey, 1987). Eddowes (1969) has rightly asserted that the main controllable factors affecting crop production in a given environment are seeds, fertilisers and manures, mechanical power, chemical power and sequence of cropping. The main objective of this paper is to see spatial variations in level of adoption of these new farm technologies. Adoption of these innovations is determined by a host of factors. This may be characteristics of innovations and of the adopters. Misra (1968) has emphasised ten out of them. These are: geographical feasibility, cultural compatibility, technological simplicity, communicability, economic feasibility, reliability, socio-technical compatibility and availability. And as such, spatial variation in level of use of these innovations are reflections of ecology, strategy, perception, attitudes, values and socio-economic situations prevailing at the time.

In the present context, only certain factors among mentioned above have been examined. Among them, irrigation is the single most important factor contributing towards raising output. Other yield-raising technologies such as high yielding variety of seeds, fertilisers, pesticides, and improved mechanical techniques have been elaborated in the present paper. Their adoption has been converted into a common index for the ease of comparison.

About the Region

For the present purpose Madhya Pradesh State situated in the heart of the country has been selected. District (45 in number) is basic unit of the present study. The state covering an area of 443.4 thousand km^2 supports a population of about 52 lakhs, constituting 7.6 per cent of the total population of the country. More than four-fifths of the people are living in the villages and more than three-fourths of the workers are engaged in agricultural occupations. Such a heavy dependence of people and economy on farming necessitates special consideration of agriculture in the development planning. From the point of view of performance of agriculture the State is far behind several other States. Yields of crops are low, dragging down the productivity. For instance, value of 18 major crops per hectare is Rs. 803 in the State, while average for the country is Rs. 1,468. Similarly, per capita value of these major crops is Rs. 284 in the State and Rs. 307 in the country. The situation of high dependency and low performance causes poverty of peasants, which itself is major constraint of agricultural development. Poverty and use of yield-raising inputs seem to be closely inter-related. For example, proportion of irrigated area and application of fertilisers per unit of cropped area are below one-third of the national averages.

Adoption of agricultural innovation manifests the complex relationship between physical and cultural milieus. Both of these environments present diversified picture in this State. Most prominent physical features in the topography are the parallel ranges of Vindhya and Satpura with the valley of the Narmada river between them. Overlocking the Narmada valley in the north, the Vindhyan ranges mark the edge of the Malwa plateau that slopes towards north. The plateau with gentle relief is bordered by the structural escarpments of the Madhya Bharat Pathar; and the low hills of several forms developed on the ancient surface of the Bundelkhand; while land escarpments and basins extend. Again, south of the Narmada valley, hills and ranges of the Satpura present very high relief and even to the east of the Maikal range as the Deogarh hills. The Chhattisgarh basin extending south of these hills presents characteristics of plain, which is bordered by the Bastar hills in south. The agricultural production and the choice of crops are controlled by the rainfall. The State occupies intermediate position between wet area in the east and arid and semi-arid areas in the west. Average annual rainfall varies from more than 150 cm in south-eastern part to less than 75 cm in the north-eastern part of the State.

Size and ownership of operational holdings determine the use of inputs and efficiency of farms. Several yield raising technologies have been made available but they are not neutral to size; and hence, they are not within reach of every peasant. Being capital intensive the use of these technologies and inputs depends, alongwith other things, on the capacity of farmers. The farmers having large holdings can and also do afford to take risk of obtaining and using yield raising inputs. Such farmers are not only capable of producing surplus even with without modern inputs, but also have easy access to financial and political power. The poor peasants with tiny parcels of relatively poor lands are deprived of such qualifications. Consequently, they can hardly afford to have inputs involving money for their agricultural development. In Madhya Pradesh, the average size of operational holdings was 3.42 hectare in 1980-81. There were more than 6411 thousand operational holdings in this state, one-third (32.8%) of which have less than one hectare of land and comprised only 4.2 per cent of total area. In all, about more than half (54.2%) of holdings are below 4 hectares in size comprising only 21.9 per cent of area. Even without making allowances for tendency and unagricultural uses of land, these holdings cannot support the peasant family and a pair of bullocks. Many resources are wasted on marginal and small holdings and it is not possible to use yield-raising technologies efficiently on these tiny holdings. On the other, extreme 6.8 per cent of holdings with large size than 10 hectares own more than 34.0 per cent of total area. Thus most of land is concentrated with a minority. On the whole, western part of Madhya Pradesh has large size of holdings and consequently higher level of adoption of development measures than the eastern part (C.K. Jain, 1988 b).

Irrigation

Among the inputs essential for agricultural development, water is one. Its use is conditioned by several variables. While low rainfall and its variable nature necessitate the development of artificial means of moisture supply, the increasing employment of fertilisers and, to some extent, of improved variety of seeds make timely supply of water as a pre-requisite. One the other side, availability of water resources, socio-economic conditions and nature of terrain affect the supply side. These variables cause temporal and spatial variation in the phenomenon of irrigation (Sharma and Jain, 1980). In Madhya Pradesh only 13.9 per cent of the total cropped area is irrigated, while country's average was 30.1 per cent in 1982-83. Similarly, in adoption of other development measures the position of the state is not much better. It is presumed that basic among the development measures necessary for the transformation of agriculture in this state is irrigation, others are very closely associated with it.

Spatial Pattern of Irrigated Area

The proportion of irrigated area varies markedly in different parts of the State as exhibited in Fig. 8.1-A. It ranges between 1.9 per cent in Bastar district to 42.5 per cent of total irrigated area in Morena district. There are 17 out of 45 districts in all recording higher percentage than the State average (13.9%). Among them, two tracts with comparatively high and very high percentages come out very clearly. One is the Chhattisgarh basin encompassing most of the areas of Raipur, Bilaspur, Durg and also Balaghat districts. This is to be noted that it is the 'rice bowl' of the State and rainfall is mostly more than 120 cm. The second tract of highly irrigated area extends over Gwalior, Bhind, Morena, Shivpuri, Datia, Tikamgarh and Chhatarpur districts. These districts with markedly low rainfall predominate in such rabi crops as wheat and jowar. Mandsaur, Ujjain, Khargone, Dewas and Hoshangabad districts of Malwa Plateau also have higher proportion of irrigated area (Table 8.1).

TABLE 8.1: M.P.: proportion of irrigated area, 1984-85

Level of Irrigated Area	% irrigated Area	No. of Districts	Districts
Very High	Above 27.8	5	Raipur, Balaghat, Tikamgarh, Morena, Gwalior.
High	20.85 — 27.8	6	Durg, Chhatarpur, Bhind, Datia, Indore, Hoshangabad.
Moderate	13.37 — 20.85	6	Bilaspur, Shivpuri, Mandsaur, Ujjain, Dewas, Khargone.
Low	10.42 — 13.90	8	Sehore, Bhopal, Shajapur, Khandwa, Dhar, Ratlam, Chhindwara, Betul.
Very Low	6.95 — 10.42	7	Rajnandagaon, Seoni, Narsinghpur, Panna, Jabalpur, Raisen, Rajgarh.
Extremely Low	Below 6.95	13	Bastar, Raigarh, Shahdol, Surguja, Mandla, Rewa, Sidhi, Satna, Damoh, Sagar, Vidisha, Guna, Jhabua.

The eastern part of the State but Chhattisgarh, is least irrigated. In Bastar (1.9%), Surguja (3.0%), Mandla (2.3%), Sidhi (3.7%) and Shahdol (2.2%) irrigation has not started in real sense. In short, the Bastar plateau, Baghelkhand plateau, Maikal plateau, Rewa plateau and the Vidisha plateau fail to have even such basic inputs as irrigation. Nevertheless, the western part is nearer to the state average than the eastern part. In these districts negligible development of irrigation is due to the undulating and highly dissected topography covered with dense forests, deep and sporadic location of sizable acquifers exploitation of which is beyond the economic means of the individuals, and low investment capability of the farmers due to their small and less productive farms.

There seems to be direct correlation of irrigated area and nature of terrain and thereby fertility of soils. By virtue of comparatively level lands and fertile fine grained loamy soils in the Chhattisgarh, parts of the Madhya Bharat plateau and the Bundelkhand uplands, it is not only easy but economical also to develop means of irrigation. Another remarkable feature of the distribution of irrigated area is the concentration in relatively more productive regions. The correlation coefficient of the value of crops with the irrigated area is direct and highly significant (r = 0.63). It has also high correlation with two important yield raising inputs, viz., use of high yielding variety of seeds (r = 0.77) and fertilisers (r = 0.64).

In Madhya Pradesh, canal is the major source providing water for 42.1 per cent of irrigated area, but it is large scale adventure and it is out of the reach of individuals. It is almost single significant source of irrigation in the Chhattisgarh and in Chambal valley. Next to canal is well irrigation covering 42.7 per cent of irrigated area. Well irrigation is popular in Madhya Bharat Pathar, eastern Satpura, Orsha upland and central Malwa plateau. Tank irrigation provides water for only 6.8 per cent of net irrigated area. It is significant only in Chhattisgarh region. Other important districts are Sagar, Seoni and Bastar.

Out of the total irrigated area more than 70.0 per cent are under only two crops — Wheat and rice. Other beneficiaries are horticultural crops, gram, cotton and sugarcane. There is marked regionalism in irrigated crops. In the irrigated tract of the Chhattisgarh basin, rice is only major irrigated crop. In other parts, though wheat is first ranking irrigated crop, other crops are also significant, such as barley and gram in the Bundelkhand uplands and Madhya Bharat plateau, horticultural crops, sugarcane and cotton in the western Malwa and in the Narmada valley.

High Yielding Varieties (HYV) of Seeds

High yielding varieties of seed technology is a revolutionary transition from age-old tradition to innovation. Farmers get more profit if they increase their output per unit area by using high-yielding varieties, chemical fertilisers and plant protection measures with ensured timely supply of water. The correlation coefficient of value of crops per unit with the use of high yielding variety of seeds is very high (r = 0.85). It is also high with the extent of irrigated area (r = 0.78). Both of these coefficient indicate that improved high yielding varieties have been adopted mostly in areas where provision of irrigation exists and where natural fertility of land is generally high (Jain, 1988 a).

The 'Green Revolution' was primarily based on high yielding variety of seeds. In Madhya Pradesh, several dwarf varieties of wheat as well as rice were evolved and introduced. By the end of the 1966-67, area under high yielding varieties was 57.7 thousand hectares in this state. Maize was first with 2.7 per cent of its area under HYV followed by rice and wheat. The area under high yielding varieties increased about 76.4 times, from 57.7 thousand hectares in 1966-67 to 4409.3 thousand hectares in 1984-85 (Table 8.2).

TABLE 8.2: Area under HYV-1966-85

Year		Rice	Wheat	Jowar	Maize	Bajra	Total
1966-67	A	32.6	8.1	0.6	16.1	0.3	57.7
	P	0.8	0.4	—	2.7	0.1	
1974-75	A	90.0	724.8	209.9	50.2	23.3	1908.2
	P	19.9	26.0	10.0	7.6	12.9	
1977-78	A	1699.8	1051.8	387.9	27.6	24.8	3191.9
	P	35.9	29.6	20.0	4.0	14.9	
1980-81	A	1502.8	1000.0	445.0	125.4	48.9	3122.1
	P	30.9	29.7	18.9	16.1	24.6	
1984-85	A	1751.1	1598.2	762.4	209.6	88.0	4409.3
	P	35.3	44.4	39.9	24.6	34.3	

A = Area in thousand hectares.
P = Percentage area of the crop under HYVs.
Source: Directorate of Agriculture, Agricultural Statistics, Bhopal, 1985.

Rice — the first ranking crop of the state, has largest (1751.1 thousand hectares) area under HYV but it constitutes only 35.3 per cent of the total area under rice. Of wheat cropped area (1598.2 thousand hectares) less than half (44.4%) was under HYV. Jowar, the third ranking crop in the state, has 39.9 per cent of area under HYV. Maize is at fourth place in respect of its area (209.6 thousand hectares) followed by Bajra (88.0 thousand hectares).

The Fig. 8.1B — and Table 8.3 depicting the proportion of cereal cropped area under HYV, show its concentration in six tracts viz., the Chhattisgarh basin — Raipur, Durg, Bilaspur, Rajgarh, Surguja and Balaghat; the Chambal sub-humid area — Morena, Gwalior, Bhind and Datia; Orchha plain — Tikamgarh and Chhatarpur; north-western Malwa — Mandsaur, Ratlam, Ujjain, Indore, Shajapur and Dewas; the Narmada valley — Khargone, Khandwa, Hoshangabad and Narsinghpur and the Satpura plateau — Chhindwara and Betul. A comparison of Figs. 8.1-A and 8.1-B shows that the districts with high proportion of irrigated area also have high proportion of cereal's area under HYV.

TABLE 8.3: Areas of major concentration of HYVs

Level of adoption	Category (in per cent)	No. of Districts	Districts
Very High	Above — 57.6	5	Morena, Gwalior, Mandsaur, Ratlam, Narsinghpur.
High	48.0 — 57.6	6	Bhind, Indore, Dewas, Hoshangabad, Tikamgarh, Balaghat.
Moderate	38.4 — 48.0	13	Datia, Chhatarpur, Shajapur, Ujjain, Khargone, Khandwa, Betul, Chhindwara, Surguja, Bilaspur, Raipur, Durg, Raigarh.
Low	28.8 — 38.4	11	Shivpuri, Dhar, Sehore, Bhopal, Raisen, Damoh, Jabalpur, Seoni, Rewa, Mandla, Rajanandgaon.
Very Low	19.2 — 28.8	3	Sagar, Satna, Shahdol.
Extremely Low	Below 19.2	7	Jhabua, Rajgarh, Guna, Vidisha, Panna, Sidhi, Bastar.

There are 21 districts out of 45 with lower proportion of cereal's area sown with HYV, forming a compact block and thwarting the south-eastern and north-western areas of high concentration. Seven of them have as low as below fifty per cent and three between 50 to 75 per cent of the State average (38.4%). These are the districts of low productivity per unit area as well as per agricultural worker.

In 1984-85, 1751 thousand hectares of rice cropped land was covered with improved seeds. There are two areas of concentration of high yielding varieties of rice. The Chhattisgarh region is most important, comprising of Raipur, Bilaspur, Durg, Rajnandgaon, Balaghat and Surguja districts. Another area of concentration is the Chambal region. It is new and minor rice producing area. As far as area under improved varieties is concerned, wheat stands at second place with 1598.2 thousand hectares under it in 1984-85. There are four areas of concentration. Most important of them is the Narmada valley stretching from Narsinghpur to Khargone. Second area of concentration is the north-western Malwa comprising five districts (Mandsaur, Ratlam, Ujjain, Shajapur and Jhabua). The Chambal region including parts of Bundelkhand uplands constitutes third area of concentration. There are six districts (Morena, Gwalior, Bhind, Datia, Tikamgarh and Chhatarpur) in this tract. Fourth area is the Chhattisgarh region including Balaghat, Bilaspur, Raipur and Durg. Area under wheat is comparatively small in these districts.

Improved varieties of maize are popular in eight districts of the western Malwa and the lower Narmada valley. They are Sehore, Dewas, Mandsaur, Indore, Chhindwara, Seoni, Ujjain and Rewa districts. Jowar the third ranking crop in the state, has 39.9 per cent of its area under HYVs. Whole of the Malwa plateau is most significant area sowing improved seeds of jowar. The Narmada valley and the Satpura region are also important. In 1984-85, there are 88.0 thousand hectares of bajra cropped land sown with improved seeds. Chambal valley (Morena and Bhind) and three districts of Malwa plateau (Dhar, Jhabua and Khargone) are most significant areas sowing improved seeds of bajra.

Chemical Fertilisers

Chemical fertilisers play a crucial role in agricultural production. Consumption of fertilisers is affected by a number of factors, the most important of which are unusual and unfavourable weather, non-availability of water, input-output relationship, non-availability of the right type of fertilisers at the place away from rail-heads and markets, absence of transport facilities, inadequacy or absence of credit facilities, lack of extension services and sale promotion campaigns and above all the ignorance of a good part of cultivators about the advantage and use of fertilisers in obtaining increased production and more income (Vishwambharan, 1982).

Manures can be organic or inorganic. Both types are necessary for good agriculture. The availability of organic manure is dependent on the extent to which green manuring crops can be grown in the fields, in addition to the making of compost and farmyard manure. The availability of farmyard manure and compost is limited as compared with the needs of cultivated crops. The cultivation of green manuring crops has to fit in with the crop pattern and is possible under irrigated conditions. As their cultivation competes with the second cash crops, it is difficult to persuade the cultivators to take up the raising of green manuring crops in a big way. The only way by which the manurial requirement can be met effectively, therefore, lies in the application of chemical fertilisers to the crops. Soil tests indicate that almost all soils in India must have additional nitrogen; about 85.0 per cent of the soils need additional phosphorus and nearly 63 per cent of them require potash, if crop yields are to be increased by 50 — 100 per cent (Randhawa, 1986).

Use of fertilisers to replenish fertility of the soil cannot be said as a popular measure in this state. On an average, 19.50 kg. of fertilisers (NPK) are applied per hectare of cropped area in this state which is more than one-third of the average for the country (46.3%) in 1984-85. In 1956-57, total fertiliser consumption in the state was 2.96 thousand tonnes which increased to 201.25 thousand tonnes in 1980-81 and by 1984-85 it reached to 372.60 thousand tonnes (Table 8.4). Total consumption of fertilisers in terms of nutrients in the state was 2960 tonnes, which included 2350 tonnes nitrogenous and 610 tonnes phosphatic fertilisers. Use of potassic fertilisers was nil. In last three decades the consumption of fertilisers increased 125.9 times and the proportion of nitrogen, phosphate and potash become 60.0, 33.5 and 6.5 per cent, respectively in 1984-85. Looking to the consumption of fertilisers per unit of cropped area, it may be observed that in 1956-57 the consumption of fertilisers was 0.16 kg/ha which went up to 9.25 in 1980-81 and 19.51 in 1984-85. It is very low when compared with national average of 46.3 kg/ha And comparison with Punjab (151.2), Tamilnadu (100.0), Andhra Pradesh (75.0) and Uttar Pradesh ([illegible].1 kg/ha) shows that the state stood very low in respect of consumption of fertilisers per unit of cropped area.

TABLE 8.4: Consumption of fertilisers, 1956-85

(000 Tonnes)

Year	Nitrogen	Phosphate	Potash	Total	Kg/ha			
					N	P	K	Total
1956-57	2.35	0.61	—	2.96	0.13	0.03	—	0.16
1965-66	14.38	7.36	—	21.94	0.82	0.41	—	1.23
1975-76	77.04	29.98	6.10	113.12	3.61	1.40	0.29	5.30
1980-81	124.95	59.88	16.42	201.25	5.72	2.79	0.74	0.25
1984-85	223.62	124.69	24.29	372.60	11.42	6.91	1.18	19.51

Source: Directorate of Agriculture, Madhya Pradesh Agricultural Statistics, 1985.

Spatial Variation in Use of Fertilisers

Application of fertilisers per hectare of cropped area in 1984-85 varies from 2.45 kg in Mandla to 56.68 in Morena. Morena, Bhind and Tikamgarh districts normally have used more than double of the state average of 19.5 kg. In the Nimar plain, western Malwa and part of the Chhattisgarh basin it is 1 to 2 times of the State average. Gwalior, Bhind, Rewa, Jabalpur and Satna districts belong to higher medium class outside of these tracts. The examination of Fig. 8.1C reveals that all districts in the Chambal region and the Chhattisgarh area with the high proportion of irrigated area have high use of chemical fertilisers. But the districts of central Malwa, Rewa-Satna plateau and Narmada valley, despite of their low developed water resource have higher use of fertilisers, naturally in dry conditions; and may be getting lower returns than the irrigated districts.

Districts with weak base come out very clearly by this analysis. The eastern Malwa, part of Bundelkhand uplands, Satpura region and Bastar plateau have low and very low amount of fertilisers used for crops. It is to note that these districts form low productivity region of the state (Sharma and Jain, 1985). It means that natural fertility of land is low and very low and no substantial effort has been made to enhance soil fertility with the application of such inputs as fertilisers and irrigation water.

There are wide seasonal variations in fertiliser consumption in the state, with bulk of fertilisers being consumed in rabi season (51.0%). It may appear that with 63.0 per cent of gross cropped area during Kharif season, the consumption of fertilisers in that season is only 49.0 per cent of the total fertilisers consumption.

Use of fertilisers per hectare of cropped area has low correlation (r = 0.39) with the value of crops per unit of cropped area, but fairly moderate with the use of high yielding variety of seeds (r = 0.51) and high with the extent of irrigated area (r = 0.64). It is evident that chemical fertilisers are mostly popular in irrigated tracts but they are not tagged with High Yielding Variety of seeds. The high positive correlation between the uses of fertilisers and high yielding variety is due to the third variable i.e. irrigation.

The fertiliser situation in the state is not satisfactory and claims for multifacet action not only to boost up consumption but also to promote balanced use and to increase fertiliser use efficiency. The constraints on fertiliser use have to be minimised by arranging cheap, easy and timely availability of credit and by supplies of fertilisers at the right time, right place and right quantity.

Plant Protection

When crop yields are increased through the use of fertilisers, good seed and improved agricultural practices, it becomes essential that they receive the benefits of protection against insects, nemotodes, viruses, fungi and bacteria. Such a protection should be given to the growing crops and to their grain during storage (Randhawa, 1986). The plant protection programme includes application of seed treatment, crop treatment, weed control, rat control and safe storage techniques. The plant protection programme covered 52.82 lakh hectares cropped area which is 23.6 per cent of total cropped area in the state in 1984-85. The seed treatment covered 20.22 lakh hectares, the crop-treatment 22.00 lakh hectares, weed control 3.46 lakh hectares and rat control 7.24 lakh hectares area in the state. The data related to the spatial aspects of the plant protection programme in the state are not available. On average, there are 5.0 plant protection machines per thousand hectare of cropped area. The high density (more than state average – 5.0) of machine use is found in nine districts, which are mostly found in Malwa plateau and the Chhattisgarh basin. In other districts, the use of plant protection machines is small and even negligible due to the limited economic means of farmers, inadequate extension services to provide training to them regarding the use of chemicals and the lack of organisations who carry out pest control operations.

Farm Implements

There are many farm implements being used here. A brief account of source of them is given below.

Iron Plough

The number of iron plough was 91.25 thousands in 1967-68 which has gone to 223.35 thousands in 1984-85 showing an increase of 144.8 per cent. Use of iron plough is very localised phenomenon. Density of iron plough varies from 0.06/000 hectare in Bastar to 88.2 in Morena and 114.8 in Gwalior districts. The mean for the state is 9.96. There are three well marked tracts, comprising 13 districts in all with higher density of iron plough per thousand hectares than the mean for the state. The sub-humid area of Chambal, consisting of Gwalior (114.8), Morena (88.2), Bhind (72.0), Shivpuri (25.8) and Datia (11.8), is most significant tract. Indore, Khandwa and Khargone districts are also significant. Damoh, Panna, Sagar and Raisen form the third area of high density. Concentration of iron plough in these districts are due to light texture of the soil. But districts of red soil zone have even lighter soils but use of iron plough is almost negligible there. There are 10 districts with density lower than the one plough per thousand hectare of cropped area, while 4 districts have less than 0.50. Most of these districts are in the eastern and southern parts of the state. Probably it is due to poor breeds of bulls employed for traction in these areas. Further, the Kharif crop dominated tracts also do not require deep ploughing for which heavy iron plough is needed. One fact comes out very clearly that people of northwest are more prone to adopt improved technology than the south-east.

Diesel Oil Pumps

Till the end of 1960 oil engine pumps were more common in this state. The number of oil engine pumps per thousand hectares of cropped area was 0.50 as compared to 0.11 electric pumps in 1960-61. Number of oil pumps increased markedly during 1967-85. Their number crossed over one and half lakh and their proportion to per thousand hectares of cropped area became 6.80. Overall increase of oil engine pumps during the period, 1967-68 to 1984-85 was 606.1 per cent, while the corresponding increase in the number of electric pumps was 3138.5 per cent, which indicates that, with the increase in rural electrification the electric pumps are becoming more and more popular.

On an average, there are 6.80 oil pumps per thousand hectares of cropped area. The density is as high as 19.4 in Khargone and as low as 1.22 in Bastar district in 1984-85. The districts with higher number than mean are mostly in western, and central parts. There are 23 districts having higher density of oil pumps than mean, and all of them are in western-central and central-eastern parts of the state. Six of them — Khargone (19.4), Khandwa (14.93), Dhar (13.75), Mandsaur (17.28), Bhopal (17.98) and Chhatarpur (14.71) have very high density of oil pumps (Fig. 8.2A). It is noted that in these districts gross irrigated area is also high. On the other hand, in three districts, there are not even two oil pumps for one thousand hectares of cropped area; and all are in the south-eastern part.

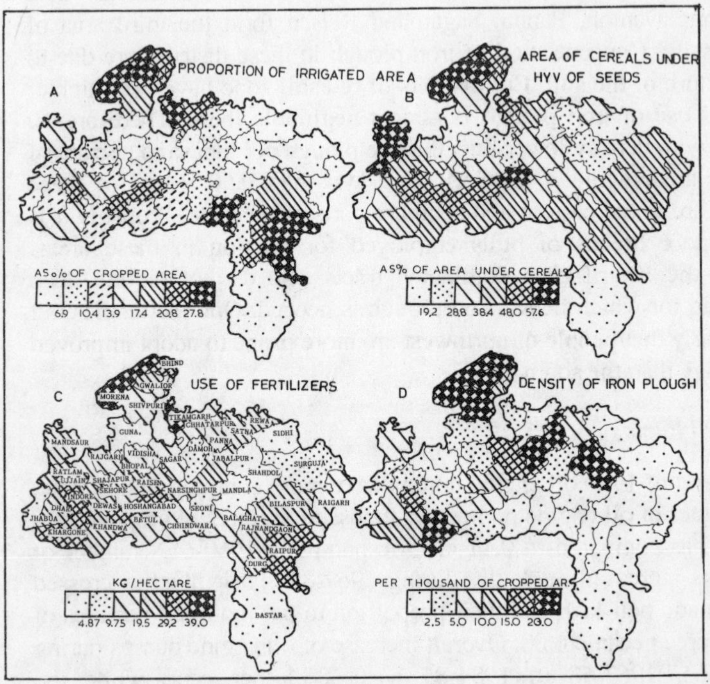

FIG. 8.1: Madhya Pradesh: Adoption of new farm technology, 1984-85: (A) Proportion of irrigated area; (B) Area of cereals under HYV of seeds; (C) Use of fertilisers; (D) Density of iron plough.

Electric Pumping Sets

Though only 52.0 per cent of villages are electrified in 1984-85, electric pump-sets outnumber the diesel pumps. There are 3,37,455 electric pumps working in the state in 1984-85 as against 1,52,646 oil pumps. This phenomenal increase is due to the electrification of rural areas. Number of electric pumps per thousand hectares of cropped area, ranges from 0.47 in Bastar to 64.75 in Indore. Again contrast between eastern and western parts of the state is evident. All 17 districts with higher density than the mean are in the western part. Only eight of them — Mandsaur, Ujjain, Indore, Khargone, Dewas, Tikamgarh, Chhatarpur and Datia have higher percentage of irrigated area than the state average. Districts with medium density are in the central part — Bhopal, Chhatarpur, Narsinghpur, Betul and Datia (Fig. 8.2-B), while deficient districts are mostly in the eastern part. Districts of the Chambal region are also poor in this respect. Sidhi, Shahdol, Surguja, Raigarh, Rajnandagaon, Raipur, Bastar, Mandla and Vidisha districts have density even lower than four pumps per thousand hectares of cropped area. Such a pattern is very closely associated with the percentage of electrified villages.

Tubewells

Tubewell irrigation based on the exploitation of ground-water has many advantages over surface irrigation by canals, as it does not involve expenditure on storage of water and is transport. It does not involve the loss through sub-mergence of large areas of valuable land under reservoirs or the disturbance of the ecological balance as is caused by big irrigation projects. The use of ground water also avoids the loss of land needed to build canal and its distributaries. It is so because private tube-wells serve only small areas of land and because the farmer who bears the cost of pumping uses just as much water as is really required for his crop and no more. From the farmer's point of view ground-water-tube-wells irrigation is superior to surface irrigation because it is entirely under his own control. A tube-well places water at the command of the farmer and he can use it as and when he is in need of it. That is why shallow tube-wells owned by farmers are more popular than deep tube-wells owned by the Government. A tube-well provides water for irrigating crops, and also for washing clothes and for bathing. Tube-wells have also saved land from water-logging (Randhawa, 1986). In Madhya Pradesh, tube-well irrigation is not much popular, due to several reasons. First and most important is the undulating terrain dominated by plateaus, hills and mountains. Others are small and very small size of land holdings, deep sub-soil water and subsistance nature of cropping. There are only 11374 tube-wells working in the state in 1984-85. Tube-wells have been successfully commissioned in the Malwa plateau and the Chhattisgarh basin. Less than half of the tube-wells (4475) working for irrigation purposes are in two districts — Indore and Dhar and they are in the Malwa plateau. Other important districts are Raipur and Durg (Chhattisgarh basin), Dewas, Narsinghpur, Ujjain, Ratlam and Raisen (Malwa Plateau). Other districts are far-behind in this respect.

Tractors

Tractors contributed to the Green Revolution, which in turn promoted the sale of tractors. In 1966, the starting year of Green Revolution, Madhya Pradesh had 2513 tractors; in 1972, the number rose to 5692; in 1984-85, it was 33,766. Tractor is comparatively much costlier affair for poor peasants of this state. Only those farmers can afford for tractor who own big land-holdings. Thus use of tractor depends mostly on size of holdings. It is predominantly small and very small in the south-east and southern parts but comparatively bigger in north, central and in north-west. This pattern of distribution of size of holdings has influenced the use of tractors.

Density of tractors per thousand sq. km. of cropped area varies from 10 in Mandla to 572 in Bhopal district. The mean of these districts' figures arrives at 150. There are 21 districts having higher density of tractors than mean, and all of them are in the central and northern parts of the state. Seven of them (Bhopal 572, Raisen 528, Vidisha 474, Sagar 407, Morena 526, Bhind 305 and Gwalior 438) have very high density of tractors (Fig. 8.2C). This is to note that proportion of agricultural labourers is low and very low in these districts; and this situation has necessitated the use of such costly implements as tractors. However, not all districts with low proportion of labourers have high density of tractors. Thus, there is not clear-cut relationship between the density of tractors and proportion of agricultural labourers. Nevertheless, districts depending on tractors are poor in labourers and flow of migrated labourers can be seen in every rabi harvesting season in these districts.

Density is low and very low in more than half of the districts which cover western and south-eastern parts of the state. It is as low as one-fourth of the state average in Bastar, Rajnandgaon, Balaghat, Mandla and Surguja in the east and in Jhabua in the west. In these two parts animal force continues to be the major source of power for agricultural operations. Some of these districts have high proportion of agricultural labours such as in southern districts, while there are others where it is low. Thus it seems that the low proportion of labourers cannot compel peasants to use labour-saving machines. It is due to several other socio-economic reasons. Alongwith the undulating terrain, small and very small size of holdings are incapable of maintaining such capital intensive implements. In fact they are not required also. On these small holdings cultivators are sufficient to perform agricultural processes. It is supported by the fact, that proportion of cultivators is usually high in these areas. Further, use of improved technology goes along with the economic development. Most of these districts are not developed and their agricultural productivity is low.

Levels of Adoption of Innovations

In the present work a comprehensive or composite index of levels of adoption of agricultural innovations is evolved through a combination of above discussed six elements. These are percentage irrigated area, use of chemical fertilisers, percentage area of cereals under HYVs, density per thousand hectares of cropped area of tractors, electric pumps and oil engines. The most crucial step in deriving such a composite index is the weighing of various elements and converting them into a common unit, so that they could be combined rationally. The procedure of indirect weighing adopted here, basically rests upon the general notion that relative weightage of unit values of various variables is inversely proportional to the total number of units of that variable. Based upon this notion the relative weightage of each variable was derived by working out the ratio between the mean sum of all variables and the actual value of variables by corresponding weightage, the weighted values of each district are summed. Taking the state average of weighted values (106.0) as standard districts are divided into six levels of adoption which are depicted on Fig. 8.2D and Table 8.5. Districts falling in each class are as follows:

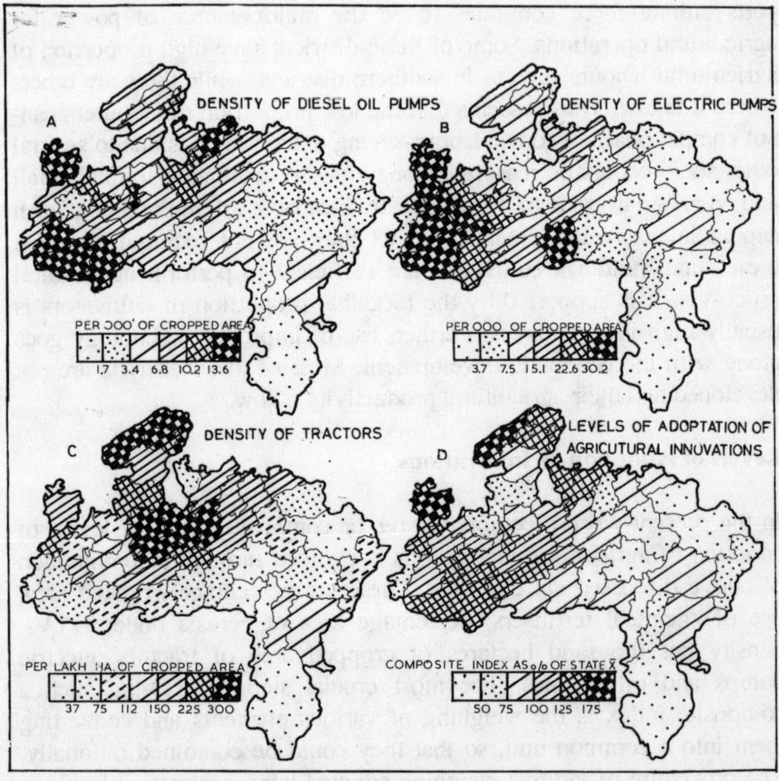

FIG. 8.2: **Madhya Pradesh: Adoption of new farm technology, 1984-85: (A) Density of diesel oil pumps; (B) Density of electric pumps; (C) Density of tractors; (D) Levels of adoptation of agricultural innovations.**

TABLE 8.5: Levels of adoption of agricultural innovations, 1984-85

Levels of Adoption	% of the State mean	No. of Districts	Districts
Very High	Above 175	2	Mandsaur, Morena.
High	125 - 175	12	Gwalior, Bhind, Datia, Tikamgarh, Chhatarpur, Hoshangabad, Khargone, Khandwa, Bhopal, Dewas, Indore, Ratlam.
Moderate	100 - 125	7	Sagar, Narsinghpur, Raisen, Sehore, Shajapur, Ujjain, Dhar.
Low	75 - 100	11	Raipur, Balaghat, Rewa, Satna, Jabalpur, Damoh, Betul, Chhindwara, Shivpuri, Bilaspur, Vidisha.
Very low	50 - 75	6	Durg, Raigarh, Panna, Guna, Rajgarh, Jhabua.
Extremely Low	Below 50	7	Bastar, Rajnandgaon, Mandla, Shahdol, Sidhi, Seoni, Surguja.

District of First Order of Adoption

Districts having scores above 175 per cent of the mean for the state are only two in number. These are Morena and Mandsaur. These districts have well-developed their water resources. It follows with high use of fertilisers and high yielding vạriety of seeds. Use of improved implements is also in vogue because of the dearth of agricultural labour. This has enhanced the productivity in this tract.

District of Second Order

These are twelve in this class, three of them are in the lower Narmada valley (Hoshangabad, Khargone and Khandwa), four in Malwa (Ratlam, Indore, Dewas, and Bhopal), three in Chambal valley area (Bhind, Gwalior and Datia), and two in Orchha plain (Tikamgarh & Chhatarpur). These districts are usually on higher side of the use of inputs. For instance, proportion of irrigated area is higher than state average in all districts but in Khandwa, Ratlam and Bhopal, which is slightly lower. Use of fertilisers is high and very high. High yielding varieties of seeds are also more popular.

Districts of Third Order

There are seven districts in this class. They cover most of the Malwa plateau (Dhar, Ujjain, Shajapur, Sehore, Raisen). Sagar and Narsinghpur are in this category. There is no uniformity among these districts, and among input even within a district. For instance, percentage of irrigated area is higher only in Ujjain but much lower in other districts. Similarly, Raisen district stands on higher and lower side in three elements considered, so this can be rated as high class of adoption.

4. *Districts of Fourth Order*

Districts of even this low class of adoption fall in the central-eastern half of the state. There are eleven such districts; two in Chhattisgarh basin (Raipur and Bilaspur), three in Satpura region (Betul, Balaghat and Chhindwara), one in the Narmada valley (Jabalpur) three in the Bundelkhand region (Damoh, Panna and Rewa), one in the Malwa plateau (Vidisha) and one in the Chambal region (Shivpuri). Use of most of inputs is low and very low. For instance, proportion of irrigated area and high yielding varieties of seeds is higher than the state average but poor in modernisation in Raipur and Bilaspur districts. Consequently, yield rates in these districts are invariably on lower side.

District of Fifth Order

Six districts fall in this class.They are distributed in Malwa plateau (Jhabua, Rajgarh, Guna), Chhattisgarh basin (Durg, Raigarh) and the Rewa plateau (Panna). These districts could not use yield-raising inputs to a great extent. However, proportion of irrigated area in Durg, density of tractors in Guna and Panna, and oil pumps in Jhabua are much higher than the state average. There is vast scope of raising productivity by increasing use of these inputs in these districts.

Districts of Lowest Order

These are poorest districts in both use of inputs and output. There are seven such districts and all are in the eastern part. Most important area is the transition belt between eastern and western parts of the state, extending from Seoni through Mandla, Shahdol, Sidhi to Surguja. Other area is formed by Bastar and Rajnandgaon districts in the rice area. Depressingly low level of use of inputs is due to several reasons. Among them unsuitable terrain, poor soils, tribal dominated population, high pressure on land and small and very small size of holdings are important. Consequently, not only present level of productivity is alarmingly low but it is declining also in some districts (Sharma and Jain, 1987). In fact, these districts should be put on priority for development planning.

Conclusion

The preceding discussion on the innovation of farm technology in Madhya Pradesh reveals a depressingly low degree of the adoption of modern farm technology. The traditional technology is yet more acceptable to the cultivators with marginal and small holdings and limited economic means which make them more tradition-oriented and powerly striken. But there is wide regional variation in level of adoption of new technology. Western and north-western districts have high and very high level. Big size of holdings, commercialisation of crops and urban influences have facilitated the modernisation of farming in this area. Contrary to it, districts of the eastern half of the state are far behind the state level. It is because of the preponderance of marginal and small holdings, high percentage of tribal and scheduled caste population, high pressure of rural population, inadequate network of institutions for supply of agricultural inputs and credit and low level of rural literacy. It is suggested that the extension agency should be geared up to spread adoption of improved farm technology. Since small and marginal farmers form a bulk of agricultural community, it is very necessary to organise and educate them for the proper choice of machinery, its proper use, maintenance and operation, so that they are able to adopt the new farm technology.

REFERENCES

Dube, R. S., 1987 : *Agricultural Geography : Issues and Application,* New Delhi : Gian Pub. House.

Eddowes, M., 1969 : *Crop Technology,* London : Hutchinson Educational, p. 61.

Jain, C. K., 1988 a : Modernisation of Agriculture and its Impact on Agricultural Productivity in Madhya Pradesh in V. Vidyanath and R. Ram Mohan Rao (eds.), *Development of India's Resource Base*, New Delhi, Gian Pub. House, pp. 398-99.

Jain, C. K., 1988 b : *Patterns of Agricultural Development in Madhya Pradesh*, New Delhi : Northern Book Centre, pp .62-63.

Misra, R. P., 1968 : *Diffusion of Agricultural Innovations*, Mysore : Institute of Development Studies, University of Mysore, p. 30.

Randhawa, M. S., 1986 : *A History of Agriculture in India*, Vol. IV, New Delhi : ICAR, pp. 291-92, 311, 254-55.

Sharma, J. S., 1982 : *Agricultural Policy in India, Growth with Equity*, Ottawa, Ont. International Development Research Centre, pp. 20-21.

Sharma, S. K. and Jain C. K., 1980 : Use of Water Resources for Irrigation and Agriculture Development in Noor Mohammad (ed.), *Perspective in Agricultural Geography*, New Delhi, Concept Pub. Company, p. 436.

———, 1985 : Changes in Agricultural Productivity in Madhya Pradesh : Spatial and Temporal Dimensions, *The Geographer*, Vol. 32, No. 2, p. 40.

———, 1987 : Green Revolution and Flattering Steps of Agriculture in the Chhattisgarh Region, *Indian Journal of Landscape System and Ecological Studies*, Vol. 10, No. 1, p. 47.

Sing, Jasbir, 1976 : *An Agricultural Geography of Haryana*, Kurukshetra, Vishal Publications, p. 187.

Vishwambharan, R. R., 1982 : Supply of Fertilisers in India, *Fertiliser Marketing News*, Vol. 43, No. 3, p. 3.

9

MECHANISATION OF SMALL-SCALE COTTON FARMS IN EAST AFRICA

DUNSTAN A. OBARA

MY FIRST challenge in dealing with the actual problem is to provide within the scope and limitations of the present paper a working definition of a small holding. However, an objective definition of the small holding is still questionable, for example, there is the problem of what standard scale to use when dealing with farm size: hectarage (ha), labour demands, capital and output. On the other hand, we find that a farm considered as a small holding in Kenya may be regarded as medium or large in terms of hectarage in another country. To avoid such confusion a small holding will be taken as a farm with less than 10-12 ha. of arable land (Coolman and Willems, 1960). Although this definition is likely to plunge us into further problems because of its subjectivity, it serves as a yardstick in the present discussion where a number of examples of small holdings cited fall within the limits of the above classification.

Since the Second World War the use of tractors has frequently been considered as the general panacea for the problems of increasing farm productivity in Africa (Anthony, et al., 1979). Moreover, mechanisation has often appeared to be the simplest and quickest way of modernising traditional agriculture, although several costly programmes for large-scale mechanisation were practised with less preparation in the 1960s. These schemes involve the purchase of large quantities of expensive equipment and a crash training programme to provide drivers and maintenance staff. But running costs have been too high and often not accompanied by correspondingly large increases in production. Despite these problems, schemes have been streamlined to manageable levels, more suitable for gaining experience on the introduction and management of tractor equipment. In Kenya, tractors were initially used by the small-farm sector for cultivation purposes in Kericho district in the early 1950s. It was here that consolidated small holdings first emerged and some of the owners purchased tractors for their own use and hire to neighbouring farmers. Today, the mechanisation of small holdings is based mainly on the hiring of tractors from both the Government and the private sector in Kenya. But the Government tractor-hire services have been provided for small-holders in Africa for at least the past two decades (Anthony, et al., 1979). The main object being to assist farmers achieve timely cultivations, but demand has heavily concentrated on land preparation, for example, in Uganda in the late 1960s about 60 per cent of the hours worked by the tractor-hire services was spent ploughing (Brown et al., 1970). Nevertheless, partial mechanisation, which only helps farmers to plant larger areas of crops, is likely to aggravate labour bottlenecks later in the season when operations are conducted manually. To obtain an efficient sequence of operations mechanisation must alleviate and not accentuate labour shortages in the small-scale farming system. Where the available farm labour is such that it limits the economic expansion of production, farmers may find it feasible to substitute capital in terms of mechanisation. If mechanisation is considered as the substitution of mechanical aids for human and animal power, then it is significant to note that small-holder cotton production in East Africa is still lagging behind a progressive mechanisation. The present paper achieves one major objective, that is, to highlight problems and opportunities of small-holder agricultural mechanisation. In order to fulfil this objective the author has cited a number of determinants of small-holder agricultural mechanisation with special reference to small-scale cotton production in East Africa. The author then provides concluding observations, policy implications of mechanisation of small-scale cotton farms and a list of references cited, which may be of interest not only to scholars but also to policy planners and policy-makers.

Determinants of Small-Holder Cotton Farm Mechanisation

The decision to introduce farm mechanisation is considered on several grounds: economic, social, political and technical (von Blanckenburg and Schultz, 1970). In addition, environmental factors play a major role in determining small-holder agricultural mechanisation. Below is a full discussion of the major determinants of mechanisation of small-holder cotton farms.

Unfavourable Quantitative Combination of Production Parameters

Economists believe that the larger the farm the easier it is to obtain optimum proportions between the production determinants: land, labour and capital. The principal problem of mechanisation on cotton small-holdings in East Africa is attributed to the fact that optimum quantitative proportions are realised only through extreme difficulty. Due to limited available hectarage, extension of arable agriculture, for example, by reclamation ranks as one of the greatest obstacles to mechanisation. In a land-abundant, labour-scarce economy it is generally accepted that capital in the form of mechanisation can be economically substituted for the relatively scarce element of production because of the reasonable expectation that mechanical cultivation will enable a farmer to expand his hectarage of tillage and that the increase in farm income resulting from this extended cultivation will more than offset the costs incurred (Clayton, 1973). However, in the land-scarce, labour-abundant countries such as East Africa mechanisation cannot help farmers to expand the size of farms. Despite this problem, mechanisation may allow a more economic combination of products. This is exemplified by the fact that the use of mechanisation to reduce labour needs at critical periods can be a key factor in development (Vail, 1973). For example, in cotton/subsistence crop farm system which is commonplace in the Lake Victoria Basin where often peak labour demand coincides with the cultivation and planting of food crops, mechanisation could prove to be economic if, by releasing labour from land preparation, it allows a rearrangement of cotton/food crop ratio, with a resulting increase in the value of output, provided the right balance is achieved between the increased revenue and the cost of mechanisation. In the case of cotton, most operations throughout the growing season are carried out chiefly by the farmer and his family. Although, farming is the sole source of income for majority of the small holders in the Lake Victoria Basin, the available labour often fluctuates in the course of time due to a number of factors: house-hold composition, migration of able-bodied people, health deteriorations due to undernourishment and diseases, traditional taboos and the like. This situation gives rise to many unused implements indicating an underemployment on the small holding. Therefore, their purchase has no justification. Underemployment on the small holdings is further enhanced because many of the small holdings are mixed enterprises consisting of two major components: livestock-raising and arable agriculture. Diversification of small-holder agriculture may be prerequisite in minimising environmental and socio-economic risks, for example, soil quality, climatic fluctuations and price changes. The quality of the soil may be maintained by a proper agricultural practice as will be seen later. In spite of the significance of small-holder agriculture diversification, mechanisation usually requires one crop to be grown at a time and it is not suited to the traditional mixed-cropping system that is often practised by small-scale cotton farmers. Cotton is often inter-cropped with beans, sorghum, maize, cassava and the like, thus rendering mechanised weeding impossible.

Farm Fragmentation

A "fragmented" farm is one whose land input comprises two or more pieces, termed "parcels", separated by land which is non-contiguous to the farmstead, most commonly being part of another farm unit. Farm fragmentation is a common feature in the Lake Victoria Cotton Area. Apart from the fact that fragmentation does not facilitate farm management, mixed-cropping, mixed farming, crop rotation, provision of credit using land as security, elimination of land disputes, and the like, the parcels are often widely dispersed, badly shaped and tiny, thus rendering machine operation all the more difficult, for example, by the longer transport distances. But due to the unfavourable farm fragmentation the savings in labour by mechanisation can be considerably minimised. The distances between the individual parcels can be covered much quicker when a tractor is employed. Despite the attractive utilisation of tractor on fragmented cotton farms, the preparation of implements for roads exerts a prohibitive influence on mechanisation. Although excessive land fragmentation has been largely eradicated through the process of land consolidation and registration in Kenya, uneconomic farms caused by population pressure and inheritance traditions remain a stumbling block to the introduction of mechanisation. A solution to this problem requires farm consolidation schemes, whereby the parcels of individual farmers are pooled together to form larger blocks situated near the homestead farm in order to facilitate more efficient mechanisation in small-holder agriculture. Block cotton farming approach has been unsuccessfully attempted in the Lake Victoria Basin where small blocks were amalgamated into larger blocks, in order to promote tractorisation of land preparation and to facilitate efficient farm management in general. While the prime cause of the failure of the mechanised cotton block cultivation schemes was inadequate organisation, mechanised cotton cultivation would appear to have contributed less to their success (Clayton, 1973). These schemes suffered from a financial failure because most of the farmers participating in the schemes experienced a fall in net income, partly as a result of the tendency to employ more hired labour on the same total cotton hectarage and partly because of low cotton yields resulting from tractor cultivation. Similar problems have been experienced in the Lake Cotton Area of Kenya. An important reason for the poor performance of group farming project in both Uganda and Tanzania was the conflict between the requirements of a participant's private family farm and his new fields in the group units, usually to the detriment of the latter (Anthony, et al., 1979). Farmers lacked the incentive to cooperate, and crop returns rarely justified the increased cost of tractor use. Similar cases have been observed in the Lake Cotton Area of Kenya where tractors have been lying idle. Moreover, the concentration of government funds and manpower on the above schemes has sometimes led to the abandonment of programmes intended to promote wider use of ox-cultivation.

Employment Problems

A study carried out in an area of high potential in Kenya of resource allocation of typical family small holdings shows that the economic justification for mechanisation and the influence of mechanisation on employment generation depend upon the ecology including the cash crops grown and the factor endowment land/labour ratio of the farms under examination (Clayton, 1970). More precisely, the results indicated that mechanisation of land preparation is most economically viable on farms in the pyrethrum zone, provided that the land-labour ratio does not fall below hectare man.$^{-1}$ At this level, hired tractor cultivation produced a profit of about 20 per cent in net income over hand-labour cultivation. Consequently, in the coffee zone it was found that income gains from mechanisation were about four per cent when the land-labour ratio was 1.08 hectares per man. These observations may assist in understanding the problems of cotton farm mechanisation in the different ecological zones of East Africa. Labour displacement by tractors is likely to be greatest in areas with suitable environment, for example, soil (Palmer, 1975). To date, the impact of mechanisation of cotton operations may be seen in terms of displacement of labour and extending the range of practicable operations. Economically, replacement of labour by machinery occurs at the periphery of profitability, that is, when a cotton farmer considers that investment in machinery, taking into account capital charges and depreciation could increase profits through reduction in labour costs (Symons, 1978). On the other hand mechanisation may be stimulated by an acute shortage of labour as has been mentioned, but the economic rule holds good because the immediate result of shortage of labour are high wages for such men and women as are available. In reality mechanisation does not alleviate labour demand, since more intensive operations may become essential, and absorb manpower on extra tasks as has already been mentioned, but the required skills change, with emphasis on mechanical knowledge replacing experience with simple implements/ tools and traction animals. To confirm this need the history of agriculture has been a succession of examples of the extension of cultivation with improved tools, from the introduction of the first light plough and its successors in heavier and more efficient ploughs and eventually, mechanically powered machines, for example tractors and aircraft. The use of aircraft for top-dressing, sowing and application of pesticides is of paramount importance in agriculture, although not simple to apply in small-scale cotton farms. It is clear that mechanisation is still lagging behind needs. Where there is a situation of rural unemployment and underemployment, social, economic and political criteria are likely to influence the decision to mechanise small-holder cotton production. When the decision is based on economic reasoning, it is assumed that the financial benefits derived from mechanisation of small-scale cotton farms will exceed the costs occurring from it. But the creation of unemployment must be viewed seriously as a cost to the entire community of small-scale cotton farms. Thus it is vital to consider farm mechanisation in relation to employment, income and other political factors. This latter problem is well documented (Inukai, 1970). The use of tractors in Uganda to alleviate labour requirements of cotton production have been considered as the best means of achieving early planting, which is crucial for obtaining high cotton yields. The data below relate to small-holder cotton production in Uganda and show that the mechanised system of cotton-growing requires less than half the total labour called for the hand-hoeing system, but harvesting requirements are similar.

TABLE 9.1: **Man-hour 0.4 hr^{-1} for mechanised and manual cultivation of cotton**

Technology	Land Preparation	Weeding	Harvesting	Total
Mechanical	11	30	210	251
Manual	136	229	223	588

Source: **McFarquhar, A.A.M. and Hall, M., 1970: "Mechanisation and Agricultural Development: no miracle in Africa", in *Options Mediterraneennes*, Paris.**

Evidence emerging from the above table indicates that mechanisation of the cotton crop in Uganda leads to a substantial reduction in employment requirements. Furthermore, the use of oxen for small-holder cotton cultivation in some parts of Uganda has been officially encouraged as being a capital-saving farm practice. But the impact on employment is no less discouraging than tractorisation.

TABLE 9.2: **Man-hour 0.4 hr^{-1} for ox-plough and hand-hoe cultivation of cotton in Uganda**

Technology	First Cultivation	Second Cultivation	First Weeding	Second Weeding	Total
Ox-ploughing	11	10	3	3	27
Hand-hoeing	50	46	46	46	188

Source: **Okai, M., 1966: *Some aspects of labour use in agriculture in the Lango district of Uganda*, Makerere University College, Kampala, RDR Paper 25.**

Again, although harvesting is not considered here, it can be concluded that the use of oxen has a substantial employment-destroying effect. Experience has shown that farm mechanisation has been uneconomic in Uganda, requiring a substantial degree of subsidisation. It has often involved rather large capital expenditure with negligible effect of the farm population. Moreover, there are no clear indications that the introduction of tractors has raised overall agricultural or labour productivity. On rather flimsy evidence, the indications are that mechanisation has reduced the labour demand, thus it destroys employment. This is the general view held by economists who argue that in economics where little structural transformation has occurred and where the absolute size of the farm labour force is increasing rapidly, investment in tractor mechanisation is likely to be uneconomic from society's viewpoint, though it is profitable to large farm operators (Johnston and Cownie, 1969). Despite these observations, there is still further discrepancy on whether mechanisation is employment-generating or employment-destroying in small-holder cotton cultivation. Inakai (1970) says that:

> Farming consists of a series of operations from ploughing the land to marketing the crops. Partial or selective mechanisation of a given operation will induce changes in the pattern of labour inputs in other operations. There is thus quite a good chance that, on balance, selective mechanisation will increase the total labour requirements of a unit of land. In other words, in a dynamic setting selective mechanisation may create rather than eliminate jobs.

This quotation which favours the idea that farm mechanisation is employment-generating is further supported by Higgs, et al. (1950) who argues that:

> Since tractors and other mechanical aids will, almost everywhere, be economic not when they displace but when they supplement manual labour, and since they will usually result in larger crop acreages to be weeded and bigger crops to be harvested, threshed and marketed, they are likely to bring about little or no unemployment . . . . Mechanisation . . . would afford employment, training and opportunities to a substantial number of native men as tractor drivers, mechanics, field supervisors, clerks, etc.

To date such contrasting views reveal the limited value of a priori reasoning for the guidance of practical policy-makers, thus painstaking empirical observations form the foundation of reaching a firm appraisal of the employment impact of mechanisation in a particular situation such as mechanisation of small-holder cotton production in East Africa.

Financing

Financing of mechanisation on the cotton small-holdings is often confronted with great difficulties. Generally speaking, the income of these farmers is not sufficiently high and stable to finance machine investments from savings. Moreover, the funds available to the cotton farm are partly determined by the composition of the household and other consumption considerations. Due to these difficulties it is of paramount importance to utilise the services of machinery contractor, or a machine cooperation either in cooperative or group use. This considerably alleviates financial constraints. In East Africa tractor has been introduced partly through the establishment of government-operated tractor hire services, and partly through spread of tractors among cooperatives, as well as individual farmers who often do contract work. Because the purchase operation, and maintenance of tractors is so costly relative to the restricted purchasing power of indigenous farmers, this innovation has affected only a relatively small number of farmers (Anthony et al., 1979).

Investment Uncertainty

It is a fact which cannot be refuted that every investment in agriculture involves risks. But the uncertainty seems to weigh more heavily for the large farms than for the small-holdings (Heady, 1952). In the case of mechanisation the enterprise becomes more vulnerable to price fluctuations. This latter situation has a major problem in small-holder cotton production, for example, in Kenya. Mechanisation reduces the elasticity of farm management insofar as it is impossible to adapt the machine park continuously to a changed cropping plan entailed by a change in price relations. Although mechanisation may allow a more economic combination of products, the costs of farm mechanisation are generally high for the small-holder with his small hectarage, particularly where he owns the tractor he operates. In this case poor maintenance leads to a short life span and limited hectarage prevents full utilisation of tractor. On the basis of this observation it seems justified that hiring tractors, especially for land preparation in small-scale cotton farms, may be more economic procedure than tractor ownership. A tractor which can be hired singly or jointly by many small-scale farmers will probably achieve greater-scale operations and the owner, for example, the Government, will possibly have a higher standard of maintenance and thus have a longer life span.

Yield Levels

The profits due to mechanical cultivation are mainly attributed to increased cotton yields, an expansion of cultivation or a more economic combination of products. For any of these to be economically viable, however, mechanisation must either reduce costs per unit of output, increase net monetary returns or reduce risks (Clayton, 1967). Due to the fact that the production factor land is in minimum position on the small holding and the factor labour is often perhaps in a maximum position the farm management generally aims at the highest possible physical production per unit of land. This intensification of farm management involves accurate land preparation, particularly in the case of heavy black cotton soils and timely operations: planting, fertilising, weeding, spraying and harvesting techniques. These factors might in certain circumstances act as impediments to mechanisation of small-scale cotton farms because the same demands for accuracy as are applied to manual labour are a prerequisite to mechanical operations. Machines which cannot meet these demands are unwarranted for small-scale cotton farms. In addition, the economic justification for tractor operation is determined by the tractor costs and the value of the higher yield. It is clear that with low-value food crops, the increase in yields must be substantial to justify the costs of mechanical cultivation. This fact seems to have been ignored in the case of many of the cooperative "tractor schemes" organised in Africa and elsewhere in the developing countries. In the case of Ugandan cotton, the impact of planting in early May instead of early June by tractor ploughing instead of hand-hoeing increased yields by about 25 per cent, thus more than offsetting the costs of tractor cultivation. Similarly, cultivation by tractor had a dramatic effect on rice yields in the Mwea Irrigation Scheme in Kenya (Gigioli, 1965). Despite these interesting observations, there is limited evidence to suggest that mechanical cultivation per se directly increases cotton yields. Improved cotton yield due to tractor cultivation is often associated with the introduction of improved cotton seeds, fertilisers, insecticides, timely planting and the like. Apart from the problems cited above, mechanisation possibly allows more intensive farm management, for example, by the growing of large hectarages of crops requiring more labour, which can then be mechanically planted, weeded and harvested. This latter task has not been possible in small-holder cotton farms in East Africa.

Environmental Determinants

(a) An obvious criterion for introducing tractors in small-scale cotton farms is the environment. In the Kano Plain and Hola tractorisation tends to be associated with irrigation, although not always. This is partly the case because land preparation is less labour-intensive on well-irrigated cotton land, say, at Hola, but this may be due to the low opportunity cost of labour in environmentally deprived areas and the returns to extra preparation of land, as well as to labour displacement by tractors in well-irrigated areas. In the Philippines, farmers planting high-yielding varieties (HYV) of rice put more efforts into ploughing and harrowing in order to incorporate weeds and straw more thoroughly into the soil (Palmer, 1975). This practice has encouraged the mechanisation of land preparation, especially in the dry season when the land is hard. Ruthenberg (1976) observed that the drier the area the more important are the interactions between mechanisation and improved varieties. He pointed out that tillage immediately after the harvest, creation of a proper soil structure, reduction of erosion, early sowings, possibly before the onset of the rains, traction power for speedy re-sowing in cases where gaps in rainfall kill the seedling, and proper and timely weeding are yield-increasing techniques which depend above all on sufficient traction power. He says that the drier the area the more pronounced are the advantages of animal traction compared with hand-hoeing and of tractors compared with ox-ploughs. Under such prevailing arid conditions mechanisation would perhaps be necessary for the success of cotton cultivation. Nevertheless, opinions differ regarding the proper approach to mechanisation in semi-arid environments. Some people feel that proper and deeper ploughing are essential, while others hold the view that minimum tillage techniques under semi-arid conditions are prerequisite. But better cotton yield, for example, at Hola has been obtained only when mechanisation is associated with fertiliser, irrigation and the like.

(b) Mechanisation destroys the whole basis of land rotation system for various reasons (Nye and Stephens, 1962). First the land must be cleared of all the large roots and stumps from which the fallow normally regenerates. Secondly, often the topsoil is thoroughly disturbed and its morphological conditions including humus content rapidly decline. Thirdly, as has already been mentioned, mechanisation usually requires monoculture, thus it is not suited to the traditional agricultural systems, for example, mixed cropping that often effectively helps in soil conservation. It is a common observation, therefore, that where mechanisation is continuously practised serious problems emerge such as squeezing pastoralists into the more marginal areas, increasing desertification, accelerated soil erosion, declining soil fertility, and increasing pernicious weeds, for example, Imperata cylindrica. The problem of soil depletion may be alleviated by the application of fertilisers or manures and crop rotation. But what is still questionable is whether fertilisers or manures combined with continuous rotational cropping and a well-designed conservation technique are alone enough or whether some form of fallowing is needed. Continuous monoculture is likely to cause problems in both the nutrient condition and the morphological condition of the soil as was the case in the Kongwa Scheme in Tanzania. In tropical soils with relatively low nutrient reserves, it is possible that nutrients may easily become unbalanced through excessive application of a particular fertiliser on the one hand and through failure to supply a deficient nutrient, perhaps a micro-nutrient, on the other (Nye and Stephens, 1962). Under normal conditions a natural fallow system maintains a reasonable balance between the nutrients added to the soil. To date, deterioration of morphological conditions is a more serious problem in forest and grassland soils in the tropics. This is partly because the organic turnover is very fast while the loss through leaching is also very fast often giving rise in some areas to sesquioxides, soils with high content of aluminium and iron oxides.

Agricultural Education

Technical education is prerequisite since the technical knowledge required to select appropriate machinery for the farm and for their technical operation and maintenance is often lacking. Similar problems of this nature have been encountered in areas of block cotton farming in Kenya (Leonard, 1977). In order to make full use of the possibilities of mechanisation it is important that agricultural education gives adequate attention to farm management. The fact that mechanisation brings none or only a small saving of labour costs implies that it will pay only when farm management is reorganised in such a way that the use of machines will result in a higher output as a consequence of the reorganised farm management, for example, a more intensive cultivation and cropping plan is essential. The foregoing argument probably gives a clue to the failure of tractorisation in small-holder cotton farms in Western Kenya where cotton output per hectare is extremely low by any standard to provide any profit to the farm, thus it might not be worth investing in tractor hire until agronomic practices are adequately implemented.

Concluding Observations and Policy Implications

Most cultivation in small-scale farms in East Africa is still carried out manually, but seasonal shortages of labour, particularly at the time of crop establishment, early weeding, and harvest limit the area of cotton crop and the quality of crop husbandry. Inadequate weeding is a common phenomenon resulting in lower cotton yields. A farmer's ability to adopt new techniques is determined by the possibilities open to him for alleviating labour shortages. A wide range of machinery is available for small farms, particularly ox-plough and spraying equipment (Boshoff, 1968; Labrousse, 1968; Minto and Westley, 1975; Darrow and Pan, 1976), but information is not readily accessible to extension services on the types of equipment that are available and on the suitability of equipment for specific tasks, socio-economic and environmental conditions. In view of these needs, it is strongly recommended that research on mechanisation problems should embrace interdisciplinary terms responsible for evaluating the need for specialised equipment and the probable consequences of mechanisation as well as determining the most suitable equipment. Subsequently, emphasis should be placed on machinery that is relatively uncomplicated and, as far as possible, capable of local manufacture, for example, ox-ploughs, small tractors and the like. However, the rate at which such equipment is introduced should be geared to a real need. It must be supported by the provision of adequate servicing facilities and the means for farmers to acquire the necessary expertise to enable them to manage their machinery to the best advantage. Nevertheless, sharply soaring fuel costs and higher prices for tractor equipment underscore that necessity of a cautious approach to tractorisation and may stimulate use of ox-plough. In order to have a thorough understanding of the effectiveness of mechanisation and its impact on income and employment, it is important to recommend that the scale of charges for tractor hire both government and private should be modified to reflect the real costs to society of using these resources. There is need for investigating and establishing the exact costs and benefits of small-holder cotton farm mechanisation in East Africa. Presently, tractor utilisation does not appear to be affected by labour-saving or cost-saving motives so much as by the opportunity to make rapid and timely land preparation in cotton farms. With uncertainty about the arrival of the rains and the needs to plant immediately after the first rains, the desirability of timely land preparation for cotton planting is likely to encourage the use of tractors despite the high rents. So far mechanisation is mainly limited to land preparation in small-holder cotton production. Apart from the environmental determinants as well as some socio-economic factors already mentioned, small-holder cotton farm mechanisation is seriously affected by traditional farming systems, particularly intercropping with subsistence crops.

REFERENCES

Anthony et al., 1979 : *Agricultural Change in Tropical Africa*, Cornell University Press.

Boshoff, W. H., 1968 : *Farming Systems in African Mechanisation of Agriculture*, in NAS.

Brown, W. T. et al., 1970 : Implements. in *Agriculture in Uganda*, edited by J.D. Jameson, Oxford University Press.

Clayton, E. S., 1967 : "Problems of peasant farm management", in von Blanckenburg and H. D. Cremer (eds.) : *Handbook of Agriculture and Food in Developing Countries*, Vol. 1.

———, 1970 : "Agrarian Reform, Agricultural Planning and Employment in Kenya", in *International Labour Review*, Vol. 102, No. 5.

———, 1973 : Mechanisation and Employment in East African Agriculture, in *Mechanisation and Employment in Agriculture,* International Labour Office, Geneva.

Coolman, F. and Willems, H., 1960 : "Mechanisation and the Small Farm", in *Mechanisation in Agriculture*, edited by J.L. Meij, North Holland Publishing Company, Amsterdam.

Darrow, K. and Pan, R., 1976 : *Appropriate Technology Sourcebook,* 2nd edition, Stanford.

Gigioli, E. G., 1965 : "Mechanical Cultivation of Rice on the Mwea Irrigation Settlement, in *East African Agricultural and Forestry Journal*, Vol. 13, No. 3.

Heady, O. E., 1952 : *Economics of Agricultural Production and Resource Use*, Prentice-Hall Inc., Englewood Cliffs, N.J.

Higgs, J. W. Y. et al., 1950 : *Report of a Survey of Problems in Mechanisation of Agriculture in Tropical African Countries*, London, HMSO.

Inukai, I., 1970 : "Farm mechanisation output and labour input: A Case Study in Thailand", in *International Labour Review*, Vol. 101, No. 5.

Johnston, B. F. and Cownie, J., 1969 : "The seed-fertiliser revolution and labour force absorption", in *American Economic Review*, Vol. LIX, Part 4.

Labrousse, G., 1968 : *Research (in the Broadest Sense) in the Field of Tropical Farm Mechanisation (Francophone Countries)*, in NAS, 1968.

Leonard, D. K., 1977 : *Reaching the Peasant Farmer: Organisation Theory and Practice in Kenya*, University of Chicago Press.

McFarquhar, A. A. M. and Hall, M., 1970 : "Mechanisation and agricultural development: no miracle in Africa", in *Options Mediterraneennes*, Paris.

Minto, S. D. and Westley, S. B., 1975 : *Low Cost Rural Equipment Suitable for Manufacture in East Africa*, Nairobi.

Nye, P. H. and Stephens, D., 1962 : Soil Fertility, in *Agriculture and Landuse in Ghana*, edited by J. B. Wills. Oxford University Press.

Okai, M., 1966 : *Some Aspects of Labour use in the Agriculture of the Lango district of Uganda*, Makerere University College, Kampala, RDR Paper 25.

Palmer, I., 1975 : *The New Rice in the Philippines.*: United Nations Research Institute for Social Development (UNRISD), Geneva.

Ruthenberg, H., 1976 : *Farming Systems in the Tropics*, Clarendon Press, Oxford.

Symons, L., 1978 (ed.) : *Agricultural Geography*. Bell and Hyman Limited.

Vail, D. J., 1973 : "Induced Farm Innovation and Derived Scientific Research Strategy: The Choice of Techniques in Developing Small-holder Agriculture in Land Abundant Areas", *Journal of Rural Development 6*.

Von Blanckenburg, P. and Schulz, M., 1970 : "The Socio-Economic Context of Agricultural Innovation Processes", in *Zeitschrift fur auslandische Landwirtschaft*, Year 9, No. 4.

10

TECHNOLOGICAL CHANGE AND SPATIAL DIFFUSION OF AGRICULTURAL INNOVATIONS IN TRANS-GHAGHARA PLAIN*

NOOR MOHAMMAD

THE TRANS-GHAGHARA Plain extending from 26°N to 28°24′N Latitude and 81°3′E to 85°E Longitude includes the districts of Deoria, Gorakhpur and Basti of Gorakhpur division, and Gonda and Bahraich of Faizabad division. It covers an area of 3,322 sq. kms. and is bounded by the State of Bihar in the east, the kingdom of Nepal in the north, and the river Ghaghara forms its western and southern boundary. The region forms a part of the Ganga plain and slopes from north-west to south-east with a gentle gradient of 40 centimetres to a kilometre. The monotony of the plain is disturbed by river valleys and drainage lines. The Terai, the Khadar and the Central Plains are the three broad physiographic units of the region. The terai belt runs in a narrow strip in the north of the region in the foothills of the Himalayas. The khadar region runs along the rivers of Ghaghara and Rapti and in fact is their flood-plain. The Central Plains lie in the Ghaghara-Rapti Doab and Rapti-Gandak Doab (Fig. 10.1).

The region has a sub-humid climate with its well-marked seasonal rhythm of monsoon. During the winter months the average temperature is 16°C., while in summer it rises up to 35°C (Fig. 10.1). About 90 per cent of the annual rainfall is received during the rainy season and the

* This paper is a part of the report of the University Grants Commission's Research Project on "Technological Change and Spatial Diffusion of Agricultural Innovations in Trans-Ghaghara Plain" undertaken by the Author.

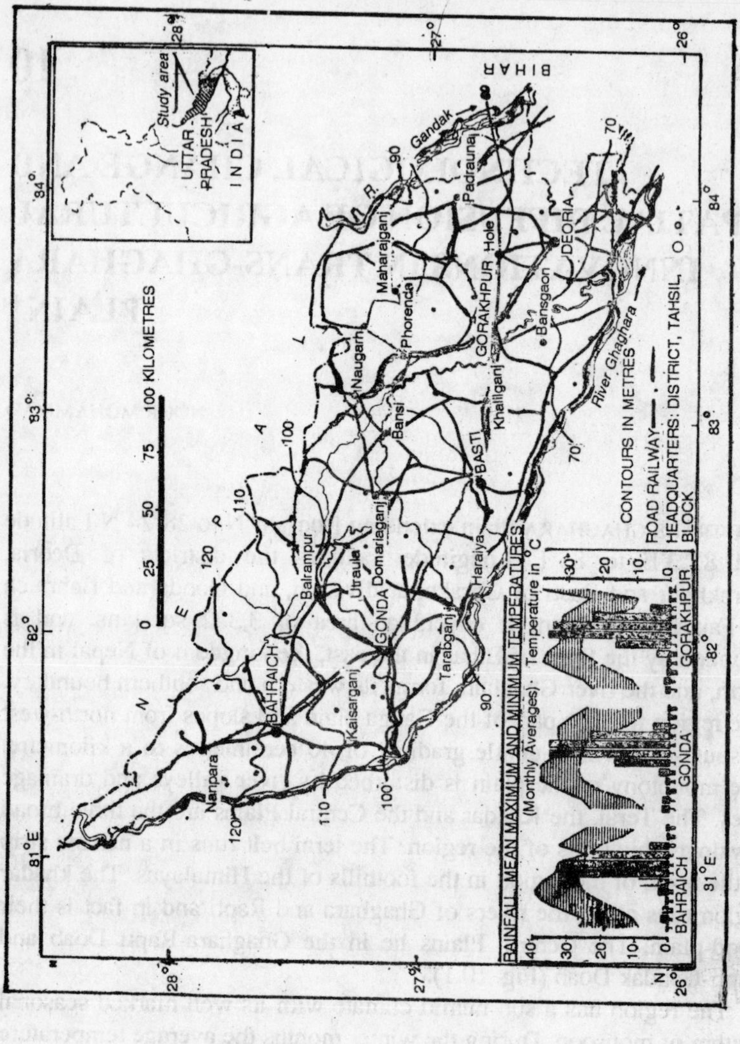

FIG. 10.1: Trans-Ghaghara Plain—geographical setting.

rest is received in the winter and summer seasons. The average amount of rainfall varies from 110 cm. in north-west to 150 cm. in north-eastern part of the region. The rainfall, though adequate for successful raising of crops, is seasonal in nature, variable in amount and its time of incidence and uneven in its spatial distribution. The variability of annual rainfall ranges between 11 and 25 per cent.

The Trans-Ghaghara Plain is made of alluvial soils which is broadly divided into new alluvium (Khadar soil) and old alluvium (bhangar soil). The former is confined to the flood plains of the river and the latter lies above the flood level. On the basis of texture and composition the soils are broadly divided into seven categories, viz. sandy, loam, sandy loam, clayey loam, clayey, *bhat* and *kachhar* (Fig. 10.2). The soil on the whole is fertile and could be put to intensive farming provided other agricultural inputs are available.

Agriculture, engaging more than three-fourths of the total population, is the main occupation of the people. Kharif and Rabi are two important agricultural seasons. The former corresponds with the rainy season, while the latter coincides with winter months (Fig. 10.2). Rice, sugarcane, maize and millets are the major crops of the Kharif season, whereas wheat, barley, peas and gram are the principal crops of the Rabi season. All the crops of the Kharif season, exceptions apart, are produced without irrigation, while almost all the Rabi crops are irrigated. The agricultural productivity is very low and the economy is also poor.

The region supports a population of 12,863,618 (1971) with a density of 387 persons per square kilometre. About 97 per cent live in the rural areas. The literary index is poor and only 16 per cent of the population is literate and educated. People are very traditional and orthodox. Social values and norms play a very vital role in decision-making. Hindus, Muslims and Harijans are the major groups. The caste composition is very complex. About 47 per cent of the population is below 20 years of age.

The size of a holding is very small. About 67 per cent of holdings are less than one hectare while two per cent come under more than five hectares. Although efforts are being made to consolidate the holdings and plots, holdings are still smaller and the size of plots is very small. The tenurial status is very complex. Big landowners are in bad shape. Irrigation facilities are inadequate and only half of the cultivated land is irrigated. The supply of electricity and diesel is very erratic. The provision of credit and subsidy is meagre. Under the procurement

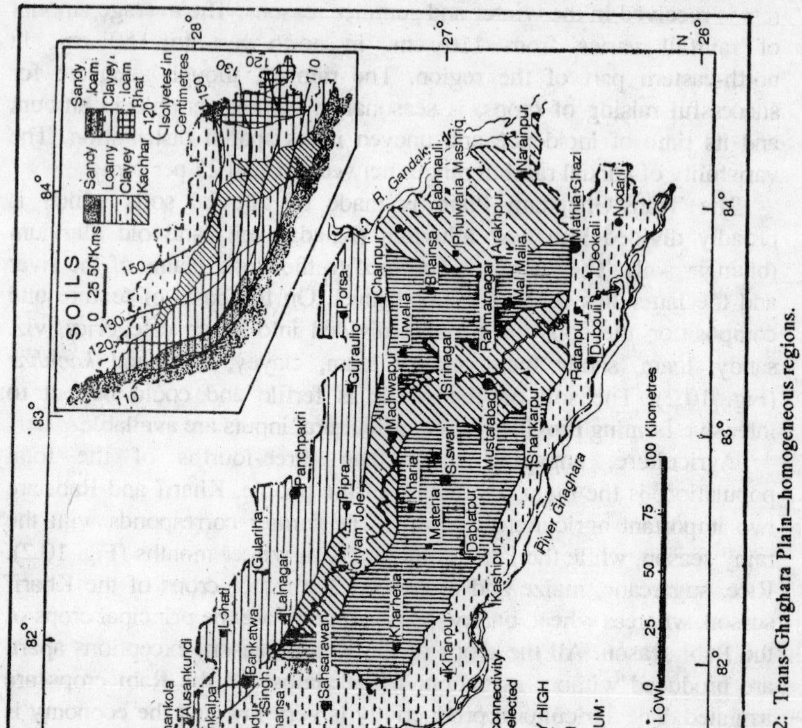

FIG. 10.2: Trans-Ghaghara Plain—homogeneous regions.

policy, the Food Corporation of India has set up its purchase centres at almost all the block headquarters, but they are not functioning smoothly and fairly. The supply of various agricultural inputs is also not properly regulated.

Sources of Data

The majority of the population lives in villages. There are 18,986 inhabited and 1,824 uninhabited villages. Hence, keeping the time and resources in view, it was not possible to conduct survey of all the villages. And, therefore, the stratified purposive sampling technique has been adopted. The region has been divided into five homogeneous strata on the basis of relief, climate, drainage, soil, landuse, cropping patterns in the Kharif and Rabi seasons together with crop combination (Fig. 10.2). From each stratum 10 villages have been selected except from Stratum V from which only five villages have been selected due to its small size. The villages within the stratum have been selected according to their communication index which have been ascertained on the basis of means of transport, their location with respect to district, tahsil, block headquarters, post office, police station, cooperative banks, educational institutions and village level workers (VLW) residence. And as such two villages from each communication strata are taken as is given in Table 10.1. As a result they differ in their physical, social and economic settings and are representatives of their region to a very large extent.

The total area of the village is 5,125 hectares of which about 75 per cent is under cultivation and less than half (48%) is double-cropped. Rice, wheat, sugarcane, maize, barley and peas are the principal crops. Agriculture engages roughly 87 per cent of the village population and the remaining 13 per cent are engaged in their caste occupation and services. Hindus and Muslims are the two major religious communities. The farmers are, by and large, orthodox and traditional. The literacy index is poor. Only 16 per cent of the total population is literate and educated which is below the State average. Literacy index of males is higher than females. However, incidentally, about 37 per cent of the sample farmers are literate.

Several agricultural innovations have been introduced in the region during the sixties and seventies. But keeping the scope of the project in view, only three major groups of innovations have been taken in the present study. They are chemical, biological and technological. The process of diffusion of innovation is influenced by social, economic and spatial systems of the region, but in the present study the impact of only economic variables has been considered.

TABLE 10.1: **Communication status of the selected villages**

Homogeneous Strata	Communication Strata				
	Very High	High	Medium	Low	Very low
I	Mal Malia Siswari	Khoria Kharbatia	Urwalia Bhainsa	Materia Daulatpur	Arakhpur Shankarpur
II	Rahamatnagar Mustafabad	Madhwapur Bhajanpur	Qyam Jote Phulwaria-Mashriq	Narainpur Bari	Srinagar Babhnauli
III	Pachpakri Bankatwa	Pipra Parsa	Gujraulia Lohati	Lalnagar Ausankundi	Gulariha Chainpur
IV	Jhansa Mathia-Ghazi	Nadauli Sahsarawan	Ratnapur Dubauli	Khanpur Deokali	Kashipur Pure Basalat
V	Sankalpa	Tawakkalpur	Bijwapur	Roshankola	P.H. Singh
Total	9	9	9	9	9

The total number of households in the selected villages is 3,748 living in 3,155 occupied residential houses. Out of this one-third households (1,249) were selected for the present study on a random basis. The field work was done in 1974-75 in both the agricultural seasons. The articulate farmers who were the functional head of the family and are instrumental in decision-making in agricultural matters were personally interviewed in order to get primary data at village and farmer levels. Besides this, district and block level data were collected from their respective headquarters.

Methodology

In the second round of the field survey, all the entries of the survey schedule were properly checked, and errors and discrepancies, if any, were removed. Weighted scores were allocated to various components of the economic system as well as to the degree of adoption of agricultural innovations. With the help of the scoring scheme the individual scores of all the respondents with respect to adoption of selected innovations were worked out. The standing score of each respondent was also calculated with respect to economic variables so as to group them in various categories. Finally, in order to know the rate of adoption of individual groups of an economic variable, the average adoption was worked out with the help of the individual adoption score of the respondents of each group. Analysis of variance and "t" tests were used in order to test the various hypotheses formulated regarding the impact of economic factors on diffusion of agricultural innovations.

Economic Factors and Diffusion of Agricultural Innovations

> Since time immemorial, thinkers were aware of the important role played by economic factors in human behaviour, social organisation, social processes and in the historical density of the society.
>
> – Sorokin, 1928

The economic factors play a very vital role in the process of diffusion of innovations especially in the case of agricultural innovations in developing countries. The classical as well as contemporary socio-psychological (Berender, 1950; Wilkening, 1950; Marsh and Coleman, 1966) writers bring out clearly the role of various elements of economic system on the adoption behaviour of human beings. During the recent times several studies have been done in our own country and abroad to assess quantitatively the impact of some of the economic indicators such as size of holdings, tenurial status, irrigation facilities etc., on adoption of agricultural innovations. The findings of these studies are varied and interesting. A group of scholars has found positive correlationship whereas the other group has got the contrary results. In some of the cases no relationship between these variables is noticed.

Keeping this in view an attempt has been made in the present study to find out the impact of economic factors (if there is any) on the adoption of agricultural innovations. The economic components chosen are size of holding, land tenure, irrigation facilities, power and diesel, credit, subsidy, input, yield and fair prices.

Size of Holding v. Adoption of Agricultural Innovations

In an agricultural economy the value of land hardly needs any emphasis and in fact it is one of the most important indicators to measure one's socio-economic status. Several sociological and socio-anthropological studies have well-established this fact (Roy, 1968; Freeman, 1961; Panse and Singh, 1966). The value of land in the region under study could be judged by the fact that even today marriages within a caste are arranged on the basis of agricultural land which is likely to come in the share of the bridegroom. However, with regard to adoption of agricultural innovations and size of holding, there are two contrasting views. According to one group of scholars there is a positive and high correlationship between the size of holding and adoption of agricultural innovations. (Mohammad, 1978; Desai, 1966; Freeman, 1961). The arguments put forth in its support are : firstly, many innovations such as tractor, power, sprayer, power duster and harvester can be used economically only in contiguous and large strips of fields; secondly, the adoption of innovation is a matter of risk-taking, because, sometimes, a widely tried innovation like IRS seed may result in a substantial failure, if the amount of rainfall is inadequate or plant protection measures are not taken timely; thirdly, many innovations require substantial amount of capital which is beyond the resources of small farmers; and fourthly, ownership of large holdings also influence the extent of adoption by way of making possible a high degree of educational status and assured irrigation facilities (Mohammad, 1978). On the other hand, there is another group of scholars who believe that there is a negative correlation between these two variables (Singh, 1965; Basu, 1978). The small farmer adopts agricultural practices earlier than the bigger one, because the former, due to low income and small production, is highly motivated to increase his level of living by adopting new farm practices, whereas the latter does not feel the need so intensely as to raise his overall output.

Besides this, he finds it difficult to concentrate his intensive efforts on an extensive area. The other view in this connection is that both variables are independent and there is hardly any relationship between the two. But the author feels that there is a positive correlationship between the two variables up to a certain size of holding (Let us say, up to 20 hectares) and beyond that the negative correlation begins.

With this view in mind the following hypothesis is formulated in order to test the relationship between the size of holdings and adoption of agricultural innovations :

> The farmers of large size of holdings tend to adopt improved agricultural practices more than the farmers of other sizes of holdings.

For the assessment of size of holding, operational holdings have been taken into account for which data were obtained during field work and were cross-checked with the records of the village accountant. The operational size of a holding varies from 0.2 hectare to 20 hectares but the average size of a holding is 1.4 hectares. On the basis of size of holding the sample farmers were classified into five groups as given below:

Group A	Large-size farmers	> 10 hectares
Group B	Medium-size farmers	10 - 4 hectares
Group C	Semi-medium size farmers	4 - 2 hectares
Group D	Small-size farmers	2 - 1 hectares
Group E	Marginal farmers	< 1 hectare

On the basis of total individual scores of the respondent, the average adoption index of each group of farmers is worked out. Analysis of variance is used to test the hypothesis. The results are given in (Table 10.2).

TABLE 10.2: Analysis of variance of size of holding

Source	Degree of Freedom	Sum of Squares	Mean Sum of Squares	'F'
Between Classes	4	386.7214	96.6804	23.4872
Within Classes	1244	5120.6772	4.1163	
Total	1248	5507.3986		

The calculated value of "F" is 23.4872 which is higher than the corresponding value in "F" test table. Hence, the test proved to be significant. As the groups differ in their average adoption significantly, our hypothesis is proved.

"T" tests are applied for all possible pairs of five groups in order to assess the significance of difference in the adoption index among the five groups of farmers. Table 10.3 shows the results. There are only 53 farmers (4.24%) whose operational size of holding is more than 10 hectares but their average adoption index is highest (3.8721). The marginal farmers (619) accounting for 49.6 per cent of the sample farmers have the lowest average adoption index (0.5421).

TABLE 10.3: **Size of holding v. average adoption index**

Sl.No.	Group	No. of Farmers	Average Adoption Index	Groups Compared	Difference in Average Adoption Index	SDE	'*t*'
1.	Group A	53	3.8721	1-2	0.4595	0.0395	11.5211
2.	Group B	118	3.4172	2-3	1.5746	0.1070	14.7225
3.	Group C	176	1.8426	1-3	2.0295	0.2441	8.3134
4.	Group D	283	0.9875	1-4	2.8846	0.3642	7.9213
5.	Group E	619	0.5421	1-5	3.3300	0.4667	7.1352
				2-4	2.4297	0.2982	8.1478
				2-5	2.8751	0.4370	6.5791
				3-4	0.8551	0.1737	4.9217
				3-5	1.3005	0.4087	3.1821
				4-5	0.4454	0.1601	2.7821

The semi-medium and medium farmers who together account for 23.54 per cent of the sample farmers have an average adoption index of 1.8426 and 3.4172 respectively. The small farmers constituting 22.66 per cent of sample farmers have an average adoption index of 0.9875. (Fig. 10.3). Table 10.3 clearly shows that there is a positive correlationship between the size of holding and average adoption index, viz. with the increase in size of holding, adoption index also increases.

Tenurial Status v. Adoption of Agricultural Innovations

Tenurial status is another important economic indicator of the status of agricultural communities. It is very vague and complex in India, generally the farmers who own land in their names enjoy a higher socio-economic status as compared to those who are either owner share-croppers or are share-croppers. The popular feeling is that owner-farmers having an assured agricultural land are more innovative and adoptive of new farm practices. But like the size of a holding there are three contrasting views with regard to tenurial status and adoption of agricultural innovations, viz. positive correlation (Savale, 1966; Bill and Zusman, 1976; Dasgupta, 1963); negative correlation (Shetty, 1966; Clive 1977; Roy, 1968); and no correlation (Chaudhary and Maharaja, 1966; Marsh and Coleman, 1966). The first view seems to be more correct as the farmers are owners of land; secondly, they need not pay to anyone if the crop fails whereas the tenant or share-croppers have to pay the contractual amount in any case; thirdly, they being assured of agricultural land can adopt permanent innovations of long duration; and fourthly, the owner-farmers are economically better off which results in a higher literacy index and more progressiveness. However, in order to test the relationship between these two variables, the following hypothesis is put forth:

> The rate of adoption among oowner-farmers is high as compared to owner-share-cropper or share-cropper.

In order to test this hypothesis the sample farmers were grouped into three categories according to their tenurial status, viz: (i) owner; (ii) owner-share-cropper; and (iii) share-cropper. For the assessment of the standing of the groups, the average adoption index of each group is calculated on the basis of the total adoption score of each farmer of the group. The analysis of variance technique is used in order to know the significance of difference between the groups. The results are given in Table 10.4.

TABLE 10.4: Analysis of variance of tenurial status

Source	Degree of Freedom	Sum of Squares	Mean Sum of Squares	'F'
Between classes	2	221.0359	110.5180	37.6193
Within classes	1246	3660.4988	2.9378	
Total	1248	3881.5347		

The calculated value of "F" is 37.6193 which is roughly eight times more than the corresponding value given in "F" test table. Hence, the test proves to be significant. Table 10.4 shows the results of "t" tests which have been applied in order to ascertain as to which group of farmers leads in adoption index and how far the three groups differ among themselves. It is quite obvious from Table 10.5 that owner-farmers who account for roughly one-third of the total respondents have the highest average adoption index (2.9274), whereas the share-croppers have an average index of 0.9312. There are only 209 (16%) exclusively share-croppers. The owner-share- croppers who account for roughly half of the total respondents also have a high average adoption index (2.1205) (Fig. 10.3) and, therefore, the analysis proves the first proposition, i.e the tenurial status and adoption of agricultural innovation are positively related.

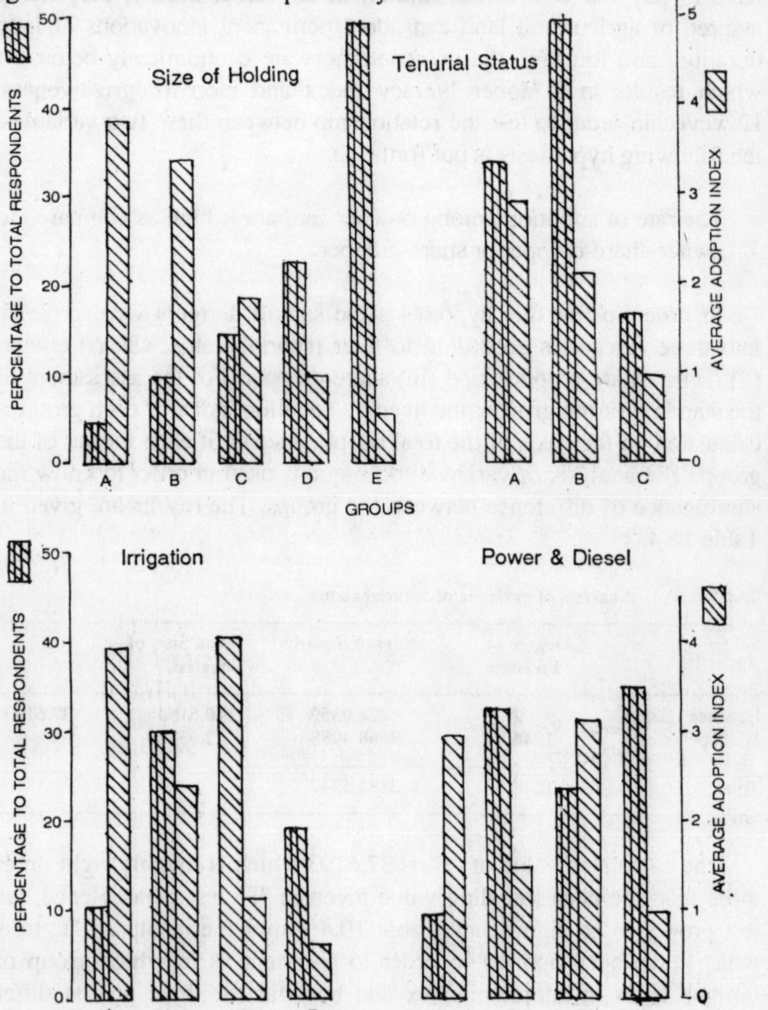

FIG. 10.3: Economic factors and average adoption index.

TABLE 10.5: **Tenurial status v. average adoption index**

Sl. No.	Group	No. of Farmers	Average Adoption Index	Groups Compared	Difference in Average Adoption Index	S.D.E.	'*t*'
1.	Owner	417	2.9274	1-2	0.8069	0.0472	17.1120
2.	Owner share-cropper	623	2.1205	2-3	1.1893	0.0902	13.1801
3.	Share-cropper	209	0.9312	1-3	1.9962	0.2201	9.0710

Irrigation v. Adoption of Agricultural Innovation

Irrigation is indeed the life-breathe of agriculture. Its importance in the development of agriculture in general and monsoonal countries in particular hardly needs any emphasis. Very often it plays a decisive role in the selection of crops to be sown, cropping pattern, intensity of cropping, crop combination, extent of yield and time of sowing the crops. A number of analytical studies have proved that India can increase its agricultural production to a large extent, if adequate and greatly assured irrigation facilities are available. In addition to this, assured irrigation encourages the adoption of certain agricultural innovations like chemical fertilisers and manures, new varieties of seeds, plant protection chemicals, because these innovations require assured and high doses of irrigation water.

Almost all the studies conducted to find out the impact of irrigation facilities on the adoption of agricultural innovations have concluded the positive correlationship between the two variables, viz. farmers with adequate and assured irrigation facilities adopt improved agricultural practices much earlier as compared to others (Mohammad, 1976; Rangaswamy, 1972; Bowden, 1965). However, in the present study an attempt has been made to test this relationship further. Therefore, the following hypothesis has been formulated in this connection :

> The farmers having adequate and assured irrigation facilities are more adoptive of agricultural innovations as compared to others.

For the assessment of the individual standing of the farmers with respect to irrigation facilities, a scoring scheme is evolved. For each crop the farmer having adequate, average, poor and very poor irrigation facilities is given the score of 4, 3, 2, and 1 respectively. His total score is worked out by adding his scores in respect of all the crops grown by him and this figure is divided by the number of crops cultivated by him in order to get his average scores. The average adoption score varies from 0 to 4. Finally, on the basis of their average scores the farmers are classified into the following four groups :

Group A	Adequate irrigation facilities	> 3.0 score
Group B	Average irrigation facilities	3.0 — 2.0 score
Group C	Poor irrigation facilities	2.1 — 1.0 score
Group D	Very poor irrigation facilities	< 1.0 score

With the help of the individual adoption score of the farmers, average adoption index of each group is calculated and finally the patterns of adoption of different groups were compared by making use of analysis of variance. The results are given in Table 10.6 which show that the calculated "F" value (19.8345) is greater than the table value and it cannot be due to random errors. Thus the test proved to be significant and the groups differ in their average adoption indices significantly.

TABLE 10.6: Analysis of variance of irrigation

Sources	Degree of Freedom	Sum of Squares	Mean Sum of Squares	'F'
Between Classes	3	237.2583	79.0861	19.8345
Within Classes	1245	4964.1885	3.9873	
Total	1248	5201.4468		

Further '*t*' tests have been used in order to understand more clearly the adoption of different groups and the difference among them. The results are given in Table 10.7 which shows that the difference is significant at one per cent level in all cases except in case of groups 3 and 4. Table 10.7 further shows that only 10 per cent of the total farmers have adequate irrigation facilities and their average adoption index is the highest (3.9871). The farmers having average irrigation facilities account for 30 per cent of the sample and their average adoption index is 2.4176. Farmers having poor and very poor irrigation facilities account for about 60 per cent of the respondents and their average adoption indices are poor, i.e. 1.1243 and 0.5871 respectively (Fig. 10.3). The above analysis is very much in conformity with the earlier studies and it proves that there is a positive correlation between these two variables.

TABLE 10.7: Irrigation v. average adoption index

Sl. No.	Group	No. of Farmers	Average Adoption Index	Groups Compared	Difference in Average Adoption Index	S.D.E.	't'
1.	Adequate	127	3.9871	1-2	1.5695	0.1339	11.7208
2.	Average	376	2.4176	2-3	1.2933	1.1564	8.2710
3.	Poor	514	1.1243	1-3	2.8628	0.5493	5.2118
4.	Very Poor	232	0.5871	1-4	3.4000	0.4715	7.2110
				2-4	1.8305	0.9146	2.0015
				3-4	0.5314	0.5314	1.0110

Electric Power and Diesel v. Adoption of Agricultural Innovations

The electric power and diesel (hereafter called power) play a very significant role in the development of agriculture especially in the mechanisation and rationalisation of farm operations (Barnea, 1953-54). Power is used in various agricultural operations such as lifting of water from tube-wells, tanks and rivers, ploughing, spraying of plant protection chemicals, harvesting, threshing and crushing. The electric pumps and oil engine pump sets are widely used for lifting water for irrigation purposes. Hence, it is expected that when power is available in abundance for irrigation and non-irrigation purposes, the farmers are in a better position to adopt agricultural innovations. Therefore in the present study the hypothesis is that:

> "Adequate and timely availability of power tends to promote speedy adoption of agricultural innovations among farmers."

On the basis of mode of consumption of power, farmers are classified into four groups. Group A consists of those farmers who use power for both irrigation and other agricultural operations; Group B includes the farmers using power for irrigation only; Group C consists of farmers using power for agricultural operations except irrigation;

and Group D is the group of farmers who do not use power at all.

The average adoption index of each group is calculated with the help of individual adoption score. Analysis of variance is used for comparing the average adoption of the groups. The results are given in Table 10.8 which indicates that F value is much higher than the table value. Hence, it is proved that the difference between the groups is significant.

TABLE 10.8: **Analysis of variance of power**

Sources	Degree of Freedom	Sum of Squares	Mean Sum of Squares	'F'
Between Classes	3	560.6241	186.8747	59.2745
Within Classes	1245	3925.1115	3.1527	
Total	1248	4485.7356		

In order to understand still more clearly the adoption of different categories, and the differences among them, "*t*" tests are applied and the results are given in Table 10.9. It is obvious that the test is significant at one per cent level in all the cases. Table 10.9 further shows that there are 437 farmers (35%) who do not use power and consequently their average adoption index is the lowest. The farmers of Group C have the highest average adoption index (3.1267) and they account for roughly 23 per cent of the sample farmers. The farmers of Groups A and B together account 42 per cent of the sample farmers and their average adoption indices are 2.9672 and 1.4512 respectively (Fig. 10.3).

TABLE 10.9: **Power v. average adoption index**

Sl. No.	Group	No. of Farmers	Average Adoption Index	Groups Compared	Difference in Average Adoption Index	S.D.E.	'*t*'
1.	Group A	118	2.9672	1-2	1.5160	0.3463	4.3782
2.	Group B	405	1.4512	2-3	1.6755	0.5374	3.1177
3.	Group C	289	3.1267	1-3	0.1595	0.0567	2.8113
4.	Group D	437	0.9875	1-4	1.9877	0.3224	6.1653
				2-4	0.5637	0.1450	3.8875
				3-4	2.1392	0.2965	7.2157

Credit v. Adoption of Agricultural Innovations

Finance is one of the most important economic determinants and life blood of any enterprise, and agriculture is no exception to it. For the agricultural development in general and adoption of agricultural innovation in particular, a substantial amount of capital is required. But the Indian farmer is too poor to afford it fully. He has to borrow money at a very high rate of compound interest from various sources like middlemen, money-lender, commission agents for his day-to-day life and occasional ceremonies. They exploit him to the maximum extent. The result is that the farmer seldom if ever gets free from the clutches these agencies. It is rightly remarked that Indian farmers, by and large, are born, live and die in debt and even bequeath debt. Hence, the question hardly arises to borrow money from these sources for adoption of agricultural innovations. However, the Government and various cooperative societies are gradually trying to help them out by providing credit facilities at lower interest rates. Several studies have highlighted the fact that if adequate and timely credit at lower interest rates is available to the farmers, they would adopt agricultural innovations more quickly. Keeping the above facts in view it is hypothesised:

> Availability of adequate and timely credit facilities promote quick adoption of agricultural innovations.

There are 781 farmers (62.5%) who had applied for credit and the remaining 37.5 per cent did not do so. Out of this a small percentage do not need credit at all but the large percentage of farmers did not apply for credit due to a number of reasons, viz. fear of high compound interest, wrong higher entries, inability to return in time resulting in social insult and harassment and uncertainty about getting it sanctioned within time. Hence, only 781 farmers have been included in the study.

In order to assess the standing of farmers in the case of credit facilities a scoring scheme in which 4, 3, 2 and 1 scores were given to adequate-timely, inadequate-timely, adequate-untimely and inadequate-untimely availability of credit respectively was used. Zero score was given for non-availability of credit. According to the scoring scheme the total score of the individual farmers was worked out and was divided by the number of items for which he had applied for credit in cash or kind. This gives his average score which varies from zero to four. According to the average score the farmers were classified into five groups as given below :

Group A	Got credit adequate-timely	> 3.0 score
Group B	Got credit inadequate-timely	3.0 — 2.0 score
Group C	Got credit adequate-untimely	2.1 — 1.0 score
Group D	Got credit inadequate-untimely	< 1.0 score
Group E	Did not get credit	0 score

The average adoption index of each group was worked out with the help of the adoption score of each farmer of the group. Finally an attempt has been made to compare the pattern of adoption of different groups by making use of analysis of variance. The results are given in Table 10.10 which shows that the calculated value of "F" is 6.7485 which is higher than the table value. Thus the test proved to be significant and groups differed in their adoption.

TABLE 10.10: Analysis of variance of credit

Sources	Degree of Freedom	Sum of Squares	Mean Sum of Squares	'F'
Between Classes	4	139.3808	34.8452	6.7485
Within Classes	776	4006.7984	5.1634	
Total	780	4146.1792		

In order to know as to which group of farmers is more adoptive and how far the groups differ among themselves "*t*" tests have been applied for all possible pairs of groups. The results are given in Table 10.11 which indicates that the test is significant at one per cent level in all cases except in the case of Groups 3 and 4. Table 10.11 further shows that there is a close relationship between the adequate and timely availability of credit and adoption of agricultural innovations. The farmers who got adequate and timely credit have the highest average adoption index (3.4107) while the adoption index of those farmers who were in need of credit and applied for that but did not get it is the lowest. There are 213 such respondents. The farmers of Group B account for the highest percentage (28.0) and their average adoption index is reasonable. The farmers of Categories C and D who together account for 35 per cent have average adoption indices of 2.9312 and

1.8796 respectively (Fig. 10.4). The above analysis proves the fact that adequate and timely availability of credit facilities for needed potential adopters promotes adoption of agricultural innovations.

TABLE 10.11: Credit v./average adoption index

Sl. No.	Group	No. of Farmers	Average Adoption Index	Groups Compared	Difference in Average Adoption Index	S.D.E.	't'
1.	Group A	76	3.4107	1-2	0.4795	0.0553	8.6783
2.	Group B	219	2.9312	2-3	1.0516	0.2406	4.3708
3.	Group C	105	1.8796	1-3	1.5311	0.8417	3.1783
4.	Group D	168	1.2013	1-4	2.2094	0.2375	9.3013
5.	Group E	213	0.5467	1-5	2.8640	0.3530	8.1125
				2-4	1.7299	0.3384	5.1125
				2-5	2.3845	0.3879	6.1479
				3-4	0.6783	0.6110	1.1102
				3-5	1.3329	0.3822	3.4876
				4-5	0.6546	0.0926	7.0715

Subsidy v. Adoption of Agricultural Innovations

The provision of incentives in some form or the other immensely contributes to increase production and efficiency. Therefore, the necessity for application of incentive has been universally recognised in a culture where everyone is working for someone else and the fruits of one's labour do not reside directly in accomplishment but accrue in some more indirect form such as wages (Poffenberger, 1961). The use of incentives as a spur to increase production rests upon a thoroughly sound psychological foundation; where incentives fail this may be traced to a failure to understand their nature, to their misuses, or to a false conception of what can be expected from them. It is evident from available data and various studies that the incentive schemes especially those involving financial benefits work well with those in poor income groups.

However, during the last two decades incentive schemes have been introduced in the field of agriculture in India in order to promote adoption of agricultural innovations, because the Indian farmer is reluctant to adopt new farm practices due to lack of capital. The general experience is that if incentives are given at proper time and in adequate measure, they contribute substantially to enhance the practice of adoption. Therefore the hypothesis put forth in this connection is as follows :

"Availability of incentives in the form of subsidies tends to promote speedy adoption of improved agricultural practices."

In order to test this hypothesis only those sample farmers have been taken who applied for various kinds of subsidies. For the assessment of the standing of the farmers with respect to availability of subsidies a scoring scheme has been evolved in which due weightage has been given to each subsidy item depending upon its adequate-timely, inadequate-timely, adequate-untimely, inadequate-untimely and non-availability.

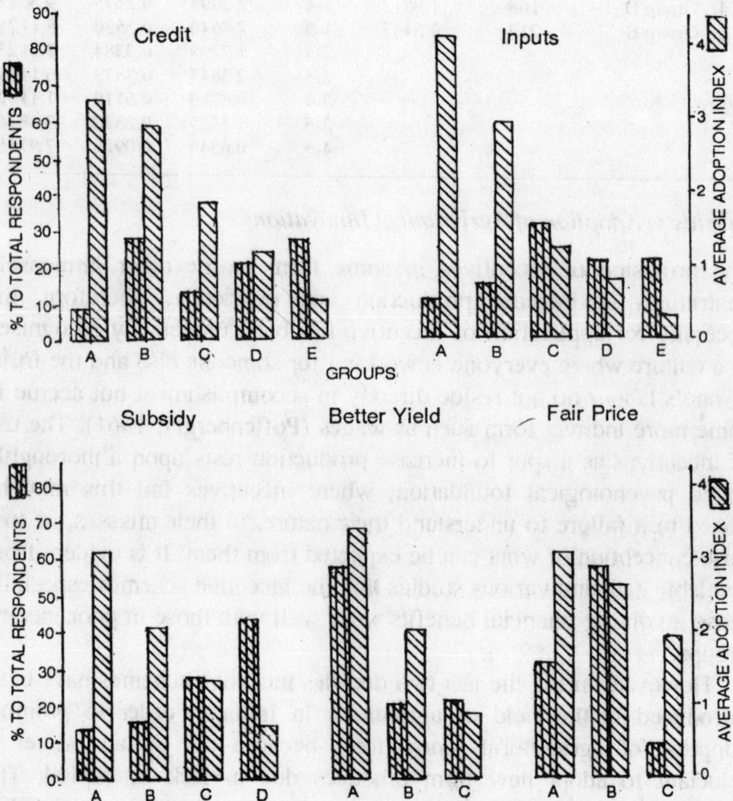

FIG. 10.4: Economic factors and average adoption index.

The average score of each farmer is worked out by dividing his total score by the number of items he had applied for subsidy. Finally the respondents were classified into following groups on the basis of their response and feeling towards subsidies :

Group A	Very good
Group B	Good
Group C	Satisfactory
Group D	Poor

The average adoption index of each group is calculated with the help of adoption score of each respondent of the group. Then the adoption behaviour of different groups was studied by comparing their average adoption index by making use of analysis of variance. The results are given in Table 10.12 which shows that the calculated value of "F" is 13.4608 which is more than the table value. Hence, the test proved to be significant.

TABLE 10.12: **Analysis of variance of subsidy**

Sources	Degree of Freedom	Sum of Squares	Mean Sum of Squares	'F'
Between Classes	3	126.7886	42.2629	13.4608
Within Classes	734	2304.5398	3.1397	
Total	737	2431.3284		

In order to probe further "t" test is applied by grouping the different classes into pairs. The results are given in Table 10.13 which shows that the test is significant at one per cent level. It is also evident from Table 10.13 that the number of respondents is in ascending order while the average adoption index is in descending order which is a clear proof of the fact that the farmers who got a very good amount of subsidies in time have the highest adoption index whereas reverse is the case for the last group.

About 13 per cent of the respondents of the study fall in Group A and 44 per cent in Group D. The respondents of Group B who account for 15 per cent have reasonable adoption index of 2.1125. The average adoption index of the respondents of Group C who account for 28 per cent is also poor (Fig. 10.4).

TABLE 10.13: Subsidy v. average adoption index

Sl. No.	Group	No. of Farmers	Average Adoption Index	Groups Compared	Difference in Average Adoption Index	S.D.E.	't'
1.	Group A	97	3.1814	1-2	1.0689	0.3630	2.9471
2.	Group B	113	2.1125	2-3	0.6912	0.1678	4.1182
3.	Group C	205	1.4213	1-3	1.7601	0.3029	5.8112
4.	Group D	323	0.7465	1-4	2.3349	0.6274	3.7214
				2-4	1.3660	0.2788	4.8997
				3-4	0.6748	0.2337	2.8871

Input v. Adoption of Agricultural Innovations

The process of adoption of new farm practices is a complex one and it involves many stages such as awareness, interest, evaluation, trial and finally adoption. It requires knowledge, skill and will among the farmers in order to make use of improved methods of agriculture. In addition to this, adequate and timely availability of various agricultural inputs like chemical fertilisers, improved varieties of seeds, plant protection chemicals and credit facilities also play a major role in diffusion of agricultural innovations. In several studies it has been observed that non-availability of adequate agricultural inputs is one of the bottlenecks in bringing about technological changes in agriculture (Ashakant, 1968). Very often inadequate and untimely supply of inputs results in cooling down of the enthusiasm of farmers in adopting an innovation. Hence, agricultural inputs should be made available to the farmer as soon as he makes up his mind to adopt them. Hence, it is interesting to study the impact of this aspect on the innovativeness of farmers with reference to some selected agricultural innovations. The hypothesis formulated in this connection is as follows :

> Timely and adequate availability of input promotes adoption of agricultural innovations.

In order to assess the standing of farmers in case of availability of inputs a scoring scheme is evolved in which adequate-timely, inadequate-timely, adequate-untimely and inadequate-untimely availability of inputs are given 3.5, 2.5, 2.0 and 1.0 scores respectively. Zero score was given for non-availability of inputs. On the basis of this scoring scheme the farmer's total score is worked out and is divided by the number of items he had applied for in order to get his average score. The average score of farmers varies from 0 to 3. Finally, on the basis of the average score, the farmers were classified into five groups as given below:

Group A	Adequate-timely	> 2.0 score
Group B	Inadequate-timely	2.0 — 1.5 score
Group C	Adequate-untimely	1.4 — 1.0 score
Group D	Inadequate-untimely	0.9 — 0.5 score
Group E	Did not get all	< 0.5 score

The average adoption index is worked out on the basis of the adoption score of the individuals. In order to test the hypothesis the average adoption indices of different groups are compared and analysis of variance technique is applied. The results are given in Table 10.14 which shows that the calculation is much higher than the table value of "F" and as such the test proved to be significant.

TABLE 10.14: Analysis of variance of input

Sources	Degree of Freedom	Sum of Squares	Mean Sum of Squares	'F'
Between Classes	4	220.6331	55.1583	17.4651
Within Classes	1244	3928.8008	3.1582	
Total	1248	4149.4339		

In order to evaluate the differences more clearly 't' test is carried out by pairing the different groups. The results are given in Table 10.15 which shows that all the groups differed at one per cent significant level except the pairs of 1-5 and 1-4. Table 10.15 further shows that the adoption index of group A is much higher (4.1465) than the rest of the groups, but there are only 139 farmers (11.13%) who have got inputs adequate and timely to a large extent. About one-fifth of the farmers did not get the required inputs and, perhaps, as a result their average adoption index is the lowest (0.3403). The adoption index of farmers of Group D who account for more than one-fifth of the total farmers is also low (0.8721).

TABLE 10.15: Input v. average adoption index

Sl. No.	Group	No. of Farmers	Average Adoption Index	Groups Compared	Difference in Average Adoption Index	S.D.E.	't'
1.	Category A	139	4.1465	1-2	1.1638	0.1186	8.4301
2.	Category B	187	2.9827	2-3	1.7367	0.3320	5.2310
3.	Category C	394	1.2460	1-3	2.9005	1.4412	2.0125
4.	Category D	272	0.8721	1-4	3.2744	2.8313	1.1565
5.	Category E	257	0.3403	1-5	3.8062	7.4999	0.5075
				2-4	2.1106	0.6087	5.4673
				2-5	2.6424	0.9142	2.8903
				3-4	0.3739	0.0908	4.1172
				3-5	0.9057	0.3127	2.8962
				4-5	0.5318	0.1293	4.1137

The farmers of Category C which constitute slightly less than one-third of the total respondents is 1.2460, while the adoption score of the remaining 15 per cent respondents falling in Group B is 2.9827 which is almost satisfactory (Fig. 10.4). This analysis clearly establishes the fact that adequate and timely availability of inputs affects the adoption of agricultural innovations to a considerable extent.

Yield v. Adoption of Agricultural Innovation

Yield is another economic variable which affects the adoption of agricultural innovations. It is altogether a different variable as it is a practice-centred characteristic whereas other variables are all predominantly far-centred. However, the agricultural innovations to be adopted by the farmer must prove superior to those it has to replace. The superiority is to be looked at from the economic as well other points of view. However, as the ultimate target of a farmer in adopting a farm practice is to get a higher return, agricultural innovations which give a quick, better and more return diffuse much earlier than those which are inferior in yield. This view has been propounded by Fliegel and Kilvin (1966) and duly endorsed by Desai (1966) in his studies. Hence, the hypothesis in this connection is as follows :

The rate of adoption of an innovation is positively related to its ultimate returns.

In order to assess the standing of farmers with respect to yield position, a scoring scheme is evolved in which a comparison could be made between per-hectare yield obtained by applying the purely traditional method and that obtained by using modern techniques. With the help of this the standing score of each farmer is worked out and they are divided into the following three groups :

Group A	Respondents getting greater yield	> 1.5 score
Group B	Respondents getting equal yield	1.5 — 0.5 score
Group C	Respondents getting lesser yield	< 0.5 score

The average adoption of each group is calculated with the help of individual scores of respondents of each group. To test the hypothesis the average adoption index of each group is compared and analysis of variance used. The results are given in Table 10.16 which shows that the calculated value of "F" is much higher than "F" value given in "F" test table. And as such the test proved to be significant. To understand still more clearly the patterns of adoption of different groups and the difference among them "t" test is used and the results are given in Table 10.16. It shows that the test is significant at one per cent level in two cases and is insignificant in the other case.

TABLE 10.16: Analysis of variance of yield

Sources	Degree of Freedom	Sum of Squares	Mean Sum of Squares	'F'
Between Classes	2	393.7807	196.8903	43.5627
Within Classes	1246	5631.5462	4.5197	
Total	1248	6025.3269		

Table 10.17 further shows that about 58 per cent farmers got good and more yield and their average adoption index is high — 3.4762. There are 210 respondents (22%) whose average adoption index is 0.9385, whereas the average adoption index of the remaining 20 per cent respondents is 2.1241 (Fig. 10.4). This analysis clearly proves the fact that the respondents who have got greater yields are more adoptive as compared to others.

TABLE 10.17: **Yield v. average adoption index**

Sl. No.	Group	No. of Farmers	Average Adoption Index	Groups Compared	Difference in Average Adoption Index	S.D.E.	
1.	Group A	563	3.4762	1-2	1.3521	0.4709	2.8712
2.	Group B	197	2.1241	2-3	0.1856	0.0745	2.4912
3.	Group C	210	0.9385	1-3	2.5377	1.7771	0.5627

Fair Prices v. Adoption of Agricultural Innovations

The availability of fair prices for the agricultural produce is another important economic indicator which stimulates the farmers to adopt agricultural innovations. It is but natural that if a farmer is sure of getting a reasonable price for his produce, he would feel encouraged and enthusiastic to increase it by adopting more and more agricultural innovations. But reverse would be the case if he does not get reasonable price for his farm product. In fact the downward pressure on price, especially when transport is deficient and storage is inadequate, may in fact be so reverse as to have a distinctive effect on producers. Premature discouragement may lead to a slowing down in good production or even rejection of a new technology. According to Wharton, unless adequate attention is given to developing of incentives, the spread of new technology may in fact be cut short before any takeoff has occurred (Wharton, 1969). Keeping these facts in view the following hypothesis is framed:

> Availability of incentives in the form of assured fair prices for agricultural produce tends to promote speedy adoption of agricultural innovations.

For the assessment of the standing of farmers with respect to availability of fair prices, a scoring scheme is introduced for all the farmers who sold their farm produce at fair prices or otherwise. There are 1,070 such respondents. Their adoption score is worked out by dividing their total score by the number of produce taken into consideration. On the basis of this respondents were classified into three groups, viz. Group A consisting of those who got good prices; Group B, satisfactory prices; and Group C, unsatisfactory prices. The average adoption index of each group is calculated with the help of the individual score of respondents. Then the pattern of adoption of different groups of farmers was studied. The average adoption indices of groups were compared by making use of analysis of variance. The results are given in Table 10.18 which shows that the "F" value is much higher than the table value. Hence, the groups differ significantly in adoption.

TABLE 10.18: **Analysis of variance of fair prices**

Sources	Degree of Freedom	Sum of Squares	Mean Sum of Squares	'F'
Between Classes	2	152.0620	76.0310	37.4205
Within Classes	1246	2531.6228	2.0318	
Total	1248	2683.6848		

In order to study it further the adoption of different groups and the difference among them, "t" test is applied and the results are given in Table 10.19 which shows that the difference is significant at one per cent level in the case of the first two pairs whereas in case of the last pair it is not significant.

TABLE 10.19: **Fair prices v. average adoption index**

Sl. No.	Group	No. of Farmers	Average Adoption Index	Groups Compared	Difference in Average Adoption Index	S.D.E.	't'
1.	Good	343	3.1245	1-2	0.4500	0.0723	6.2193
2.	Satisfactory	627	2.6745	2-3	0.6918	0.1735	3.8976
3.	Unsatisfactory	105	1.9827	1-3	1.1418	1.9730	0.5787

Table 10.19 further shows that the largest percentage of the respondents (58) are satisfied with the price they have got for their agricultural produce and their average adoption index is also high. There are 343 respondents who have got good prices whose average adoption index is the highest (3.1245). Only 10 per cent of the respondents were not satisfied with the prices they got and their average adoption index is low (Fig. 10.4). The above analysis proves

that the availability of fair prices for agricultural produce does promote speedy adoption of agricultural innovations.

Conclusion and Suggestions

It goes without saying that the present study is too brief to make any generalisation either at global or at national level with regard to the relationship between the technological change and spatial diffusion of agricultural innovations and economic factors, but predictions could be made with a fairly high degree of certainty in this connection for all the geographical regions with similar environmental, social and economic conditions. It is quite obvious from this study that economic factors play a significant role in the process of diffusion of agricultural innovations. All the nine economic factors considered in the present study have proved statistically significant and all have positive correlationship with diffusion of agricultural innovations. In the light of the above conclusions the following suggestions are made with regard to different economic variables.

The entire Suryupar plain is almost covered by the settlement scheme of consolidation of plots and in most of the sample villages this programme is almost over. But the size of the plots is still, by and large, very small. The size of the plots could be judged by the fact that about 67 per cent of the holdings are less than 0.5 hectare. It is due to several reasons : firstly, the consolidation programme is not properly and rationally chalked out. There are several flaws in it. It is not properly implemented as well and the farmers are given more than one plot in the same type of land. The farmers are harassed and they have to pay heavy prices for it directly and indirectly. Secondly, farmers themselves prefer to have at least one plot in each soil type due to subsistence economy. Thirdly, the social tradition of Indian farmers to distribute the land among their sons and grandsons has further reduced the size of plots. Therefore, it is recommended that a rational programme of consolidation should be implemented and legal restrictions imposed on fragmentation of holding. Besides this, an attempt should be made to consolidate the small holdings through cooperative and joint farming.

With regard to tenurial status it has been proved that owner-farmers have the highest adoption index as compared to share-croppers. There are several reasons for this. The uncertainty in the duration of tenancy, psychological feeling of detachment, small size of operational holding and plots and consequently irrational and haphazard planning of cultivation are some of the important reasons. It is, therefore, suggested that the land should belong to its real cultivators and, if it is not possible and practicable, share-croppers should be given large-size of holdings on contract basis for longer periods of time.

Adequate and assured irrigational facilities also play a decisive role in the adoption of agricultural innovations. It is, therefore, strongly suggested that the government should give top priority to development of irrigation facilities in its plans and policies in order to revitalise Indian agricultural technology. Various kinds of subsidies should be given to the farmers in various forms to develop medium and minor irrigation schemes. Besides this, farmers should also be educated about the proper and profitable use of water as most of them do not know the ideal frequency of irrigation and optimum water requirements of the crop. Hence, the government should also encourage research on these lines so that farmers could apply irrigation water economically and cultivate their fields with the optimum use of water. Likewise power also affects the adoption process. Hence, power generation should be accelerated to cope with the increasing demand of power supply for both irrigation and other agricultural purposes. Similarly the supply of diesel for all agricultural purposes should be rationalised and the farmers assured of getting it when they need it. Both the measures would go a long way in speeding up the mechanisation of farm industry.

There is a strong and positive relationship between the prompt availability of inputs and adoption of agricultural innovations. Therefore, in order to make the same available in time and in adequate quantity, various distribution centres should be opened by government and private agencies. Similarly proper arrangements must be made to give adequate and timely credit to the farmers whenever they require it.

The study further showed a strong and positive correlation between yield and adoption of agricultural innovations. The rate of adoption of such innovations as give increased yield is much higher as compared to others. Hence, an attempt should be made to introduce and propagate such improved practices, especially the new varieties of seeds in research centres which do meet these requirements.

Needless to mention that if the farmers are assured of increased production and better gain, they will respond in large numbers to all the agricultural innovations.

Adequate and timely availability of subsidies also promotes adoption of agricultural innovations. Therefore, all kinds of incentives should be given to the farmers in the form of subsidies in cash or kind through various government and quasi-government agencies. Likewise, the surety of getting fair prices for agricultural production also promotes adoption of new farm practices. Hence, every attempt should be made to give higher prices to agricultural products through fair price shops, regulating markets, providing storage facilities. However, it is finally suggested that policy makers and social researchers should give much emphasis to economic variables.

REFERENCES

Azad, A. S. and Garg, P. L., 1977 : Family Planning and Poverty, *Mainstream* (Jan.).

Basu, A., 1978 : *Technological Possibilities of Indian Agriculture*, Calcutta, p. 41.

Bernder, R. W., 1950 : *The Influence of the Group on the Judgement of Children*, New York.

Bear, R. M., 1947 : *The Social function of Education*, pp.46-47.

Bowden, L. W., 1965 : *Diffusion of the Decision to Irrigate*, University of Chicago, Research Paper No. 97.

Bose, S. P., 1966 : *Socio-Cultural Factors in Farm Efficiency*, Department of Agriculture and Community Development, Calcutta.

———, 1965 : Peasant Values and Innovations in India, *American Journal of Sociology*, Vol. 67, pp.552-60.

Bose, S. P. and Dasgupta, S., 1962 : *The Adoption Process*, Department of Agriculture and Community Development, Calcutta.

Brow, L. A. and Moore, E. G. : *Diffusion research in Geography*, Discussion Paper No. 9, Department of Geography, Iowa University.

Chaudhuri, D. P., 1979 : *Education, Innovations and Agricultural Development*, Delhi.

Cuber, J. K., 1958 : *Sociology : A Synopsis of Principles*, New York, p.199.

Davis, K., 1956 : *Human Society*, New York.

Dube, S. C., 1955 : *Tradition, Social Structure and Agricultural Development*, Kurukshetra.

Emergy, F. E. and Oeser, O. A., 1958 : *Information Decision and Action*, Psychological Determinants of Change in Farming Techniques, Melbourne.

Farooqi, M. Y., 1972 : *Land Utilisation in Trans-Rapti Plain* (unpublished Ph.D. Dissertation submitted at A.M.U.).

Good Caster, V., (Ed), 1959 : *Dictionary of Education*, London, p.191.

Luiton, R., 1962 : Cultural and Personality Factors Affecting economic Growth in Best, H.F. (Ed), *The Progress of Under-Development Areas.*

Misra, R. P., 1968 : *Diffusion of Agricultural Innovations*, Mysore.

Mohammad, N., 1974 : "Diffusion of an Agricultural Innovation in Haryana", *The Geographer*, No. 1, Vol. XXI, Aligarh. *The Geographer*, No. 1, Vol. XXIII.

———, 1978 : "Impact of Social Factors on Diffusion of Agricultural Innovations in Central Trans-Ghaghara Plain", *The National Geographical Journal of India*, Parts 1 and 2, Vol. XXIV, Varanasi, pp.36-46.

———, 1978 : *Agricultural Landuse in India*, Delhi.

———, 1978 : Impact of Economic Factors of Diffusion of Agriculture Innovations, in Central Trans-Ghaghara Plain, *Geographical Review of India*, Calcutta.

———, 1976 : Technological Change and Diffusion of Agricultural Innovations, *The Geographer*, No. 1, Vol. XXIII, Aligarh.

Mohammad, N. and Majeed A., 1979 : "Socio-Economic Factors and Diffusion of Agricultural Innovations in Mohammad, A. (Ed), *Dynamics of Agricultural Development*, Delhi, pp.151-74.

Rahudkar, W. B., 1962 : "Farmers' Characteristics Associated with the Adoption and Diffusion of Farm Practices", *Indian Journal of Agricultural Economics*, Vol. XVII, pp.82-85.

Raj, K. N., 1977 : *Mainstream*, December.

Ramachandran, R., 1975 : *Spatial diffusion of Innovations in Rural India*, Mysore.

———, 1976 : A Survey of Spatial Communication Behaviour and Adoption of Farm Innovations, *University Grants Commission Project Report*, Delhi.

———, 1979 : "Spatial Diffusion of Innovations in Agricultural and Farmers Communication and Travel Behaviour in Meerut District, *ICSSR Project*, Report, Delhi.

Rajgopalan, C. and Singh, J., 1971 : *Adoption of Agricultural Innovations*, Delhi.

Rogers, E. M., 1958 : Categorising the Adopters of Agricultural Practices, *Rural Sociology*, Vol. 23, pp.345-54.

Sachchidanand, 1972 : *Social Dimensions of Agricultural Development*, Delhi.

Shafi, M., 1974 : Perspectives on the Measurement of Agricultural Productivity, *The Geographer*, No. 1, Vol. XXI.

Singh, A. K., 1967 : "Hindu Culture and Economic Development in India", *Conspectus*, No. 1, pp.9-32.

Singh, S. M. and Reddy, S. K., 1965 : "Adoption of Improved Agricultural Practices of Farmers", *The Indian Journal of Social Work*, Vol. XXVI.

Sorokin, P., 1928 : *Contemporary Sociological Themes*, New York, p.514.

Torsten, Hagerstrand, 1952 : *The Propagation of Innovation Waves, Lund Studies Series in Geography*, Lund.

Wharton, C. R., 1969 : The Green Revolution : Comicopia or Pandoras Box, *Foreign Affairs*, No. 3, Vol. 47, p.3.

Wilkening, E. A., 1950 : "A Socio-Psychological Approach to the Study of Acceptance of Innovations in Farming", *Rural Sociology*, pp.352-64.

Van de Ban, A. W., 1960 : Locality Group Difference in the Adoption of New Farm Practices, *Rural Sociology*, Vol. 25, pp.308-20.

11

GREEN REVOLUTION AND ITS SOCIO-ECONOMIC IMPACT ON INDIAN RURAL LANDSCAPE

SURENDRA SINGH

Green Revolution

GREEN REVOLUTION may be termed as the human struggle for survival despite obstacles and various types of constraints. It is a phase in the development of human civilisation where scientific researches have proved their social relevance. It is an aggregate of mechanical, hydrological, chemical, biological and administrative inputs in the transformation of rural economy. William Gaud, the Administrator of the U.S. Agency for International Development (USAID) used the term "Green Revolution" for the first time in a speech — "The Green Revolution : Accomplishment and Apprehensions", before the Society for International Development (SID), Washington in 1968, in which attention was focused to a major break-through in the agricultural production in the developing parts of the World. In India the Green Revolution heralded in mid-sixties (1965-67) with the introduction of high yielding dwarf varieties of Mexican Wheat and some exotic varieties of paddy on selected scale. Though it is a debatable point whether the so-called Green Revolution is really a green revolution or a type of "Agrarian Revolution/Agricultural Transformation" which is a result of some favourable measures undertaken in the developing countries after the throughing-off the yoke of imperialistic exploitation in yester-years. Because the term "Green Revolution" would imply fundamental changes in farm economy, in the inter-relationships between the farm and external world and in addition, altered social relationships on the farm and a shift of power with in the rural sector from one class to another (Parthasarthy, 1971).

The term "Green Revolution" has a global acceptance in the academic parlance and social recognition, hence, its use cannot be avoided. However, in the Indian context this term conjures in the minds of many people a vision of agricultural prosperity which is not in consonance with the reality (Singh, 1974). The emergence of Green Revolution in India may also be closely associated with land consolidation and some other land reforms. But it is not spread to all types of agro-climatic regions. For example, in the context of attaining food self-sufficiency less than 15 per cent of the area under the foodgrains in the country (mostly in the Green Revolution Belt) has contributed as much as 56 per cent of the increase in foodgrains production in the post-Green Revolution period (Dantwala, 1987). The agrarian revolution has been thus created due to integration of external flow of inputs and know-how of technology with demands for the output. The inter-action of farm communities has increased tremendously with the urban communities and the isolation of the peasantry in relation to money has been completely broken (Parthasarthy, 1971). This change has revolutionised the agrarian production and saved the country from starvation.

Green Revolution in India has changed the traditional subsistence farming system to market-oriented production system. It has resulted in multi-facet socio-economic changes in rural areas. The concept of increasing returns per unit of area and per unit of time is the basic consideration. The traditional native farming technology was replaced in Green Revolution areas in the country which has been instrumental in some structural changes in the agrarian set-up. The technological inputs in the form of tractor and associated implements, tubewells and pumping sets, electricity, threshing and winnowing machines alongwith harvest combines etc. has resulted in positive gains everywhere. Associated biological, chemical and water technology led to continuously increasing volume of harvest. Structural changes in the agrarian sector encompasses a cross-section of inputs and outputs closely related to various types of farming. Consequently the resultant influence will be different in various areas. These changes have also brought pattern of crop specialisation in India because the infrastructural development proved favourable for particular type of crops while restrictive for others, e.g., regular growth in irrigational facilities and area have restricted the growth of dry-farming crops.

Genesis of Green Revolution

Green Revolution is the net outcome of a particular type of agrarian system covering a wide variety of measures to enhance the agrarian output. This revolution is based on high yielding varieties of seeds (HYVS) with closely associated agricultural technology, manures, chemical fertilisers, insecticides, pesticides, weedicides, biological stimulators alongwith capital investment under well-guided irrigation facilities, i.e., mechanical, chemical, biological and hydrological inputs are essential for such agrarian transformation. Introduction of institutional inputs with regard to land reforms, consolidation of land holdings, banking system and credit facilities, associated service infrastructure in the form of rural link roads, rural electrification, tubewell irrigation, marketing support, storage facilities etc. have also been instrumental in bringing the Green Revolution. These institutional inputs have brought structural changes in agrarian economy in the country. Keeping all other facilities aside, the introduction and developments in the field of irrigation — tubewells and canals, have been most rewarding in the genesis of the "Green Revolution".

Spatially Green Revolution is limited to certain parts of the country where necessary infrastructural development in associated aspects is better than other regions. The most prominent among such areas are northwestern Rajasthan-Punjab-Haryana-Chandigarh-Delhi-Western Uttar Pradesh-Region of the Northern Plains; Krishna-Godavari Delta (Thanjavur Area) in Coastal Tamil Nadu and parts of Maharashtra. The Green Revolution is appreciably localised in some particular areas of toiling farming communities like the Jats (Punjab, Haryana, Western Uttar Pradesh), the Reddys (Andhra Pradesh) and the Gounders (Tamil Nadu) who have been entrepreneur cultivators since ages and they carry almost a mystical love for their farm holdings (Singh, 1974). In Punjab and Haryana besides economically viable and socially acceptable technological package, the public policies and services provided were conducive for giving a lead in the field of agricultural affluence (Chopra, 1986). Now Green Revolution is penetrating in some other areas where infrastructure is being developed. Agricultural development is a continuous process and in this process development penetrates into new areas, adjacent to the old areas or distant areas. Regional disparities have been widened in India in rural development after the Green Revolution. This has led to many socio-cultural and politico-economic repercussions in the country.

It is estimated that in Punjab-Haryana increase in agricultural production due to the consolidation of the farm holdings alone without change in the techniques was reported to be 25 per cent (Randhawa, 1974). Land consolidation not only provided compact farm holdings to cultivators but also encouraged to undertake several output enhancing measures like adopting high yielding varieties of seeds, installation of power energised tubewells for easy and required irrigation, chemical fertilisers and modern mechanical inputs etc. The consolidation of land holdings makes it possible for the individual farmers having sizable holdings to set-up their own tubewells for the assured supply of water required for the successful cultivation through a self-controlled system. Power input is also essential for increasing production. National Commission on Agriculture recommended one horse-power input per hectare to get yield of the order of over one tonne per hectare. Rural electrification in Punjab, Haryana, Western Uttar Pradesh, North Western Rajasthan and other parts of the country prove this notion that rural electrification is essential for agrarian development also.

Like land reforms, farm machinery has also been instrumental in augmenting farm production. Farm machinery is the direct form of energy input in agrarian sector. Farm machinery has resulted in several structural changes in agrarian economy in India. Tractor is the basic farm machinery which is capable of supporting other mechanical inputs. Punjab has only 1.53 per cent of the total geographic area of the country but has over 25 per cent tractors and 8 per cent of tubewells of the country. Similarly, highest ratio of stationary threshers, combined harvestors, tillers, harrows, puddlers and chemicals spraying apparatus etc. in Punjab is also proving the relationship between Green Revolution and farm mechanisation. Process of farm mechanisation has directly been instrumental in the realisation of the Green Revolution.

Increased productivity, area and production of certain crops has also influenced the market mechanism in agrarian sector. Over 60 per cent of the foodgrain production is rendered surplus to be marketed in Punjab, Haryana, parts of north-western Rajasthan and western Uttar Pradesh. Farm inputs requirements also increased tremendously with the advent of Green Revolution demanding huge investment in a traditionally static agrarian landscape. It has resulted in the stimulation of institutional infrastructural development in rural areas which indicates the consequence as well as the stimulator of agrarian development.

Technological transformation of agriculture in India has not resulted in dramatic changes and moreover its localised nature has set a chain of action for surrounding peripheral areas for agrarian reforms and development. Rural roads have also been proved to be carriers of agrarian progress and means of rural development. It has eased the transportation of inputs and marketing of outputs. Increased accessibility is also changing social attitude in rural areas and spread of education.

Green Revolution and Agrarian Development

The self-sufficiency in foodgrains in India may be ascribed to the Green Revolution. Annual foodgrain production in the country was 51 million tonnes in 1948 which touched 120 million tonnes in 1976, 150 million tonnes in 1984 and 151 million tonnes in 1986, recording an overall growth of three times in thirty seven years. Such increase in agrarian sector has never been recorded in India. It has been possible due to increase in cropping intensity. This development has not only made India self-sufficient in foodgrains but also has generated a deep rooted confidence for feeding its fast increasing population. The realisation of Green Revolution in India is primarily achieved through expansion in irrigation facilities. In 1949-50 total irrigated area in the country was a little over 19 million hectares which increased to over 63 million hectares in 1983-84 and to 70.4 million hectares in 1986-87 recording an overall growth of 3.7 times in thirty seven years. It has resulted in an allround change in endogenous and exogenous linkages of agriculture, agricultural community, village and society, breaking the isolation of the peasantry. The total estimated potential of irrigation in the country has been estimated around 113 million hectares. This increase is possible in the northern, western and southern sectors of the country and these areas have already out-run other areas in agricultural yields (Soni, 1987).

Green Revolution has been the cause of the unusual spread of rice (a wet crop) to relatively dry areas of the country notably in Haryana, Punjab and north western Rajasthan due to assured supply of water for irrigation from canals and tubewells. Similarly, wheat has penetrated in non-traditional areas of Haryana, Rajasthan, Maharashtra, Karnataka, Tamil Nadu, Andhra Pradesh, Orissa, West Bengal and Assam giving a new dimension to our agrarian landscape. The constant rise in production is the direct outcome of balanced inter-action among various favourable forces including proper inputs, encouraging policies and quick adaptability of the farm community (Singh, 1987). Green Revolution has not only changed the agrarian landscape but also helped other economic activities based on agro-products giving a new set-up to farming systems.

Foodgrain production in Punjab was 32 lakh tonnes in 1960-61 which touched the figure of 172 lakh tonnes in 1985-86 recording an overall increase of 5.4 times in 25 years resulting socio-economic changes in rural Punjab. During 1968-69 wheat recorded an increase of 229.5 per cent, rice 1221.2 per cent , sugarcane 20.9 per cent, rapeseed 80.1 per cent and mustard 126.9 per cent which is much higher than the national average except in the sugarcane. Realisation of Green Revolution came through the increased productivity of cereal crops. In 1966-67 average yield of wheat in the country was 8.8 quintals per hectare which reached to 13.8 quintals per hectare in 1971-72 and 18.2 quintals per hectare in 1982-83 recording an overall increase of about 2.1 times in sixteen years. In the corresponding period wheat yield in Punjab was 15.1 quintals per hectare in 1966-67 which rose to 24.1 quintals per hectare in 1971-72 and 30 quintals per hectare in 1982-83. In Haryana, it was 14.3 quintals per hectare in 1966-67, 20.4 quintals per hectare in 1971-72 and 25.2 quintals in 1982-83. Growth in yield in Punjab was two times and in Haryana 1.8 times. In average yield Punjab and Haryana are number one and two in India. Similarly increasing yield in other crops has also been recorded. The level of yield indirectly reveals the level on inputs for enhancing the production and also indicates the potentialities (Singh, 1987).

The phenomenal rise in area under cereals and their production may be attributed to the carefully planned inputs strategy in the field of judicious use of fertilisers, high yielding seeds and availability of water for irrigating the thirsty lands alongwith crop protection programmes, various incentives, subsidies and procurement of cereals by various government agencies to provide the economic umbrella to the farming community and avoiding the situation of glut in the market. This all has resulted in the orientation of agrarian system in the country in favour of cereal crops which has deeply influenced other crops (Singh, 1987).

Problems Associated with Green Revolution

Socio-econmic impacts of Green Revolution have been different in various areas according to geographic and socio-political environment. It has averted famine and hunger despite the brunt of population explosion in various parts of the globe including India. So it may not be improper to designate it as the foodgrain revolution. No doubt, the Green Revolution has heralded the era of self-sufficiency in the foodgrains but not without its repercussions and problems in the field of socio-cultural and economico-political fields, because the dynamism of the Green Revolution has been introduced into a tradition-bound static rural society in India. The process of change in the traditional rural societies is usually very slow and accepted values are cherished despite their outmoded existence.

Green Revolution in India was first heralded over the area which receives relatively less rainfall, i.e., the semi-dry areas of the north-western India covering Punjab, Haryana, parts of northwestern and northeastern Rajasthan and western Uttar Pradesh due to the facilities of canal and tubewell irrigation which have been very encouraging for wheat and rice cultivation, while the dry-land cultivation is gradually becoming out of date in these areas. Demand and requirement of water for crops is regularly increasing.

Agriculture is a land-based occupation which is a non-elastic and immovable resource. After the Green Revolution the significance of land holdings have tremendously increased alongwith its social value. It has also increased its demand and with the increasing productivity this demand is also rising fast. Land-holding is fast emerging as the point of social power and status. As the new techniques are not favourable for small holding farmers so such people sell their meagre holdings and become landless labourers. It has also been observed that the farmers with medium holdings (6-12 hectares) are eagre to increase their holdings. Important relationship exits between land and people who strive about the nature of property rights, distribution of ownership and control of land, system of land division and settlement patterns (Smith, 1959). Such relationships are constantly changing in the areas of Green Revolution.

Land transactions directly influence the distribution of holdings according to size. Number of large and medium sized holdings increase and small sized ones are abolished due to land transactions. Thus the position of operational land holdings' size seems to be controlled by economic and technological forces. Heralding of Green Revolution has made the traditionally popular system of tenancy cultivation unpopular and the holding owners now prefer to cultivate their holdings themselves owing to enhanced returns. Green Revolution has also adversely affected the share-cropers, tenant farmers and landless agricultural labourers because 40 per cent of the total cultivated area of the country is owned by six per cent farm households (over 8 hectares) who are economically capable to invest for Green Revolution. It has made the rich farmer richer and it has resulted in the concentration of wealth and rural power in the hands of rich farmers and the poor peasants have not benefitted at all (Dutt and Sen, 1986). Similarly, benami transfers of the land on a very large scale in these areas have also adversely affected the poor and landless farmers as the government could get very little surplus land to be distributed among them. Green Revolution has caused decrease in the number of small and medium farmers due to economic and technological constraints. It has resulted in the increase in the number of relatively large holdings either by selling or leasing of the land. It has also caused significant income differential between the income of farmers and the farm labour. In the Punjab the farmers' income had recorded an increase of about 195 per cent during 1979-80 but the increase in the income of farm labourers could rise only up to 114 per cent during the corresponding period (Arora, 1987). It has widened the economic gap in the rural areas which will have its own impact on the socio-economic stratification of the rural society. More risk bearing capacity of large farmers, their greater political power and control over the developmental resources and agencies have provided them access to credit and input supply system. While the farmers with relatively limited economic capacity languish behind in this race.

After the advent of the Green Revolution increased economic returns have encouraged the farmers to enlarge their holdings either by purchase or encroach on the common lands or reclaim the sub-standard areas or clear the fruit plantations adjacent to the towns or cities. This tendency is widely prevalent in agriculturally rich areas. Particularly cutting of fruit plantations around the settlements have devoid them of their traditional green belts. New plantations of dwarf species giving fruits early are of very restrictive utility for green belt purposes. Track gardening around the urban centres has also now become most potent factor for clearing of the plantations. Fast growth of towns and cities has also adversely influenced the fruit plantations on the outer periphery of the old parts of the towns and cities both for agricultural and colonisation purposes.

Notable changes have been recorded in the agriculture after the

Green Revolution. Agricultural labour is one of such fields which has recorded notable changes. It has also influenced the occupational structure due to the increased productivity, mechanisation and replacement of traditional farm techniques with the modern methods. Certain types of agriculture have become labour intensive and produced a unique labour cycle. Local labours have migrated to the cities and towns in the neighbourhood for more remunerative jobs or small business or self-employment keeping the residence in villages. All this had resulted in a wide gap in labour availability and the requirement. Consequently migration of labour from distant areas increased. Factor of cropping pattern and its nature also affected the supply of labour. Rice cultivation in the Punjab has increased at a very fast rate and it has penetrated to totally new and non-traditional areas even in the southern Punjab. But the local cultivator is not conversant with its cultivation. So the required labour comes from eastern Uttar Pradesh, Bihar, eastern Madhya Pradesh and parts of Orissa. According to an estimate such labour force for the crop-husbandry may be varied from four to six lakh people (Grewal, 1988). Local labour joins this work only during the peak season of sowing and harvesting when the wages are maximum in the year. Thus the traditional labour-farmer relations have changed from Jajmani relations to financial relations on the contractual basis. In some aspects of agriculture mechanisation has also affected the labour negatively resulting out in the need of labour for various farm jobs which were previously done manually causing unemployment. Growing unemployment in rural areas will have its own socio-economic consequences.

Marketing of the surplus foodgrains and other agrarian products becomes problem during the peak marketing season after harvesting because the farmers want remunerative return just after harvesting. Market is flooded with the produce occasionally creating the position of glut which causes sharp decline in prices because the development of the market infrastructure by the concerned state-owned agencies could not cope with the increasing volume of marked surplus. Mechanisation and modernisation of the farming system has not been followed by equally competent marketing and handling system.

Agrarian techniques developed in last two decades are favourable for irrigated cultivation and to a great extent dry farming is not cared while about 70 per cent of our agricultural area is unirrigated and rainfed which contributes only 40 per cent of the country's total agricultural output. Since all parts of the country are not favourably located in relation to be developed as the irrigated areas. So major thrusts have to be made in agrarian research favourable for dry farming. It will also help in tiding over the problem of unemployment prevalent in these areas. Development of efficient dryland farming techniques has become a national socio-economic necessity in India of present age. The benefits of agricultural transformation should spread to the maximum possible areas in the country and should percolate to the grass-root level. So serious efforts have to be made to develop dry farming system alongwith necessary land reforms.

Green Revolution has caused a serious disturbance in the cropping pattern in its sprawl. The traditional crop rotation has been soil fertility conserving and enriching. Fertility recharge is essential to get better yields during the next crop. But in new crop rotation one soil exhausting crop is followed by another soil exhausting crop with no intervening period of rest and fertility recharge for the soil.

Introduction of new factors in agrarian landscape in under-developed areas sometimes produces unique and unprecedented results. Introduction of canal irrigation hitherto dry areas has noticed such changes in various parts. It has even become the ultimate cause of out-migration and large scale land transactions because the farming community of the area was not aware of the irrigational know-how which attracted people from irrigated areas. Land transactions rendered local people landless labourers. Later it became the cause of social tensions and rising economic disparities which disturbed the special fabric in a peaceful society.

Consequences of Green Revolution

Long evolved traditional cropping pattern and crop rotation has been replaced by the Green Revolution by set pattern of irrigated crops and lapse of the fallow land system in most of the areas influenced by it. Notably area under the foodgrains and cash crops has increased while area under pulses and dryland farming crops has decreased in comparable manner. One crop immediately succeeds another just after the harvesting. In Punjab, Haryana and Western Uttar Pradesh wheat succeeds rice or cotton or vice versa, exerting heavy pressure on the soil fertility. This succession is tried to be maintained with the use of heavy inputs of chemical fertilisers and flood irrigation. This has resulted in the soil salinity in all these areas alongwith harmful effects on soil fertility. It has also resulted in agrarian encroachment by the wheat and rice in non-traditional areas demanding heavy energy inputs in various forms. No doubt that the scientific advances in agrarian sector have enhanced the geographic limits of different crops. But in this race the ecology is being harmed knowingly and unknowingly without any research supports. Increased productivity, economic returns and input structure created and naturally available has also affected the crop selection but continuously increasing demand of the food-stuffs is among the most potent factors for increasing the area under food crops. It has resulted in short supply of various other agro-products like barley, pulses and has influenced the area and production of oil seeds. Consequently the country has to import these items in large quantities. Moreover, the per capita consumption of the pulses and oilseeds in the country is decreasing. This all needs proper and concerted planning for crop diversification and need-based agrarian programming.

Inflow of the monetary gains in the rural areas after the Green Revolution has also changed the consumption behaviour. But expenditure on some particular items has increased disproportionately. Expenditure on food is relatively increased much less but has multiplied on clothing and liquors. In some areas consumption of liquors has increased at an alarming rate, notably in cotton and sugarcane growing areas. While expenditure on sanitation improvement and environmental regeneration is almost negligible. Similarly litigation regarding land ownership and associated cases has also increased. Development of production and productivity has resulted in demand of creation of certain amenities in rural areas, like drinking water, surfaced roads, transport services, health and sanitation facilities, regulated markets, regular supply of electricity and further electrification, creation of more storage facilities close to the villages, improved housing facilities, educational facilities and local availability of consumer goods to avoid unnecessary commutation.

Transfer of the cultivated land also testifies the fact that during the period of Green Revolution the large farmers are increasing their real assets while the marginal farmers are selling their small pieces of lands and moving to the urban areas in search of employment. Uprooting of a person who has born and brought-up in a farmer's family leads to psychological depression and social tensions (Husain, 1987). Such changes in the socio-economic senario in rural areas are also leading to changes in other sectors causing polarisation of landless labourers, marginal farmers, share-croppers and labourers unemployed due to the mechanisation of the agriculture. This polarisation is becoming a great danger to the peace and tranquility in the rural society.

Changes in economic relations and migration of labour force to the Green Revolution areas have also brought its own implications both for the migrants and locals. It has been observed that migrant labourers feel a unique type of alienation whereas the local labourers feel that it has not only adversely affected their economic prospects but also harmed their bargaining capacity for the wages with the wealthy local farmers. This feeling brings social tension which has already been created due to economic differentials after the Green Revolution. Though the economic conditions of the migrant labourers have improved considerably in comparison to their own village folk but change in their life style, dialectic tone, food habits, new skills they learnt and changed social attitudes encourage others to migrate. This creates lack of skilled farm labour in their own area.

A number of industries are dependent on agriculture for the supply of raw materials (flour mills, shelling mills, husking plants, jute, cotton, oil seeds, sugar cane, tobacco, food processing, beverages etc.). But the Green Revolution has initiated the linkage of the agriculture with the industry more progressively for the farm machinery, chemicals and fertilisers etc. Dependence of agriculture on industry has proved to be constantly capital intensive activity causing rise in production cost of agro-produce.

Technological transformation of agriculture has also given incentives for the adoption and development of many associated pursuits like dairy farming, poultry farming, vegetable cultivation, pomiculture and in some areas has encouraged rural industries. It has also enhanced the chances of developing allied economic activities in the field of flour milling, rice shelling, oil presses, farm equipment repair and financial institutions (banks, cooperatives etc.). Thus Green Revolution has been a factor of generating employment in rural sector. Increased production, early maturing crops and multiple cropping obviously created diverse job opportunities (Chibb, 1987).

Green Revolution, to this day, has only influenced the traditionally rich agricultural and better off areas of the country. Rest of the country is as under-developed as it was prior to Green Revolution. This has developed an economic landscape with unbalanced growth, socio-economic disparities and regional imbalances. Despite Green Revolution about 80 per cent of the people cannot afford to pay for balanced diet, consequently malnutrition is not unusual (Singh, 1974). Green Revolution has also a capitalising influence on Indian agriculture because it is day by day becoming cost-oriented due to high ratio of energy consumption in all allied fields. The small and poor farmer is not in a position to meet the challenge because economically he is not capable to meet the heavy investment in inputs and moreover is not capable to obtain soft loans, grants and subsidies which are grabbed by wealthy and powerful landowners in connivance with the revenue and development officials. Thus the commercial farmer has become affluent at the cost the subsistence peasants (Bhat, 1987).

At present almost no attention is being paid to the ecological impact and consequences of the Green Revolution. Use of high quantities of fertilisers, chemicals, pesticides, insecticides, weedicides, fungicides, biological stimulators and other synthetic inputs etc. have their own effects and after-effects on the environment. Effect of their residuals is also not known to the required level. Even ground water resources are influenced. Similarly environment of traditional social understanding, brotherhood, cooperation and socio-psychological security in rural areas is disturbed because of the economic gains of the Green Revolution. It has resulted due to accentuation of imbalances in overall economic development in Green Revolution areas.

Conclusion

Under sharp regional contrasts in natural environment and the resultant variation in land quality, it is possible to identify land systems for evaluation landuse potential which is measured by using capability for alternative crops and per hectare productivity. Because of these differences and variations in availability of irrigation it would not be possible to attain regional equity in growth of agricultural production and productivity. Level of regional development based on agriculture will vary in different agricultural regions according to agro-climatic set-up, soil, nature of crops, availability of irrigation and inputs, land carrying capacity, productivity and social forces. Inter-regional disparities have become more sharp than the inter-regional differences under the impact of Green Revolution.

Impulse of agrarian development has to be generated for the peripheral and distant suitable areas from central areas of present Green Revolution [Punjab north-western-Rajasthan-Haryana-Delhi-Western Uttar Pradesh Plains; Krishna-Godavari Delta; Cauvery Delta

(Thanjavur Area); Coimbatore-Madurai Area and parts of Maharashtra]. Outward expansion of these growth areas will not only relieve these areas from the pressure of foodgrain production but new potential areas will emerge and crop diversification urgently needed for conservation may be introduced.

Growing consumerism in rural areas will ultimately exert extra-ordinary pressure on the resources. So it should be very carefully kept within the economic capacity. Because consumeristic race is contrary to the traditional brotherhood found in our rural areas due to its very competitive nature of developing individualistic social set-up. The effect of negative consequences of Green Revolution may be minimised through properly planned social education. Various types of agencies engaged in rural development, agriculture, dairying, education, promoting the use of different kinds of agrarian inputs, social groups and government organisations alongwith the voluntary organisations may work in the field of social education including the eradication of the traditional evils of the rural society. To arrest the growing unemployment, restricting the widening economic gap and social tensions more opportunities for gainful employment should be created in the rural sector. Growing population and growing unemployment is one of the most important causes of social tension in the rural society. New evils growing with the economic development are also harming this tender fabric of our social system. So thoughtless mechanisation of agriculture should also be checked. For the development of India rural development is inevitable and the Green Revolution has provided the required thrust so it should be planned for creating balanced development, but not with the widening economic gap and socio-cultural problems. The cooperative movement needs to be further strengthened to create job opportunities in the rural sector. Similarly rural industrialisation and the patronisation of their products by the government and people also needs fresh assessment and planning.

REFERENCES

Arora, D. R., et al., 1987 : *Socio-economic Consequences of Green Revolution in Punjab*, Deptt. of Eco. Sociology, Punjab Agricultural University, Ludhiana, p. 4 (mimeographed).

Bhat, L. S., 1987 : *Agricultural and Socio-Economic Development — A Regional Perspective*, Indian Statistical Institute, New Delhi, p. 5. (mimeographed).

Chibb, S. S., 1987 : *Green Revolution in India*, Directorate of Correspondence Courses, Punjabi University, Chandigarh, p. 4 (mimeographed).

Chopra, P. N., 1986 : *Green Revolution in India*, New Delhi, p. 15.

Dantwala, M. L., 1987 : Growth and Equity in Agriculture, *Indian Journal of Agricultural Economics*, Vol. 42, No.2, p. 151.

Dutt, R., and Sen S., 1986 : *Indian Economy*, New Delhi, p. 502.

Govt. of India, 1976 : *Report of the National Commission on Agriculture*, Ministry of Agriculture, New Delhi, Pt.xv, p. 75.

Grewal, S. S., et al., 1988 : *An Analysis of Farm Income in Punjab*, Deptt. of Eco. and Sociology, Punjab Agricultural University, Ludhiana, p. 7. (mimeographed).

Husain, M., 1987 : Socio-Economic Consequences of Green Revolution — An Appraisal, Paper Presented at the National Conference on Dynamics of Green Revolution, *Geographic Perspectives*, Deptt. of Geography, Punjabi University, Patiala, Oct. 5-7, 1987, p. 6 (mimeographed).

Mahadev, P. D., 1988 : Tribal Area Development in Reservoir and Command Area of Irrigation Project, Paper Presented at the National Seminar On Environmental Management in Hill and Tribal Areas, Deptt. of Geog., N.E. Hill University, Shillong, Oct. 14-17,1988. (both papers mimeographed).

Mahadev, P. D. and R. N. Achyutha, 1988 : *Prosperity and Improvement in Regional Development Planning — A Case Study of Tungbhadra Command Area*, Institute of Development Studies, University of Mysore, Mysore.

Parthasarthi, G., 1971 : *Green Revolution and the Weaker Section*, Bombay, p. 1.

Randhawa, M. S., 1974 : *Green Revolution : A Case Study of Punjab*, New Delhi, p. 43.

Singh, Jasbir, 1974 : *Green Revolution In India — How Green Is It*? Vishal Pub., Kurukshetra, 40.

Singh, S., 1987 : Dynamics of Wheat Culture In Haryana, Paper Presented at the National Conference on the Dynamics of Green Revolution, *Geographic Perspectives*, Deptt. of Geography, Punjabi University, Patiala, Oct. 5-7, 1987, p. 17 (mimeographed).

Smith, T. Lynn, 1953 : *Sociology of Rural Life*, Harper, New York, p. 217.

Soni, R. N., 1987 : Seventh Five Year Plan, Suggestion Approach to Agricultural Development — A Case Study for Dry Farming. S. Bhatnagar and S.S. Chibb (Eds), *Rural Development in India, A Strategy for Socio-Economic Change*, Delhi, p. 182.

12

SOCIO-ECONOMIC IMPLICATIONS OF TECHNOLOGICAL CHANGES IN AGRICULTURE*

NOOR MOHAMMAD

Conceptual Framework

RURAL DEVELOPMENT in India is, in fact, an euphamism for the development of agriculture which continues to be the most premier industry of the country engaging roughly 70 per cent of its population directly or indirectly, and contributing nearly half of the national income. It also provides bulk of our export items, and is a major source of our industrial raw materials. Therefore, since independence, our country has decided to follow a massive induction of technology—a time tested tool to tackle any crisis situation, into the sphere of agriculture, specially in case of agricultural implements, high yielding varieties of seed, chemical fertilisers and manures, improved irrigation facilities, and insecticides and pesticides. The new farm technology began to bear fruits as early as mid-fifties but an outstanding breakthrough in production could be achieved only in late sixties when several package schemes and new strategies were adopted for the development of agriculture. As a result foodgrains production increased considerably, and recently we have become not only self-sufficient in foodgrains but have unprecedented stock of about 20 million tons.

Every technology acts as vehicle of change and modern farm

* The study is essentially based on the primary data collected by Mr. B.K. Rathore, and Mr. S.H. Qaiser for their M. Phil dissertation written under my supervision. Their permission to use the data is gratefully acknowledged.

technology is no exception to this rule. Leaving exceptions apart, improved farm technology has acted as unfailing engine to accelerate the pace of socio-economic development in our country. Compared with the general poverty of pre-independence period, the farmers of today have undoubtedly reached a stage of relative prosperity and well-being. Their standard of living, housing condition, agricultural infrastructure, household assets, degree of mobility, attitude towards the education and profession of their children, religion, culture etc. have changed and improved. In fact the new found prosperity has brought about a complete transformation in the socio-economic life of the countryside.

But the diffusion of farm technology is not uniform all over the country both in space and time, and consequently, socio-economic consequences also vary considerably. At meso-level the rich farmers have become more rich and the poor have not been able to keep pace with them. Hence, the gap between the rich and the poor has further increased. Even at micro-level, the fruit of new farm technology has been laped by the big landlords, and the poors have been further impoverished. The intriguing changes brought about the new farm technology, has opened a new vista of research and investigations. This paper is an endeavour in this direction.

The problem of rapid socio-economic development of the countryside has long been pre-occupying the minds of policy-makers and political leadership. Now they can hardly ignore the conclusions drawn by the scholars of different discipline. A number of studies conducted so far reveal that the process of agricultural modernisation is' sped up by boosting a number of socio-economic factors — education, mass media, extension services, while the induction of farm technology brightens the socio-economic picture. Thus it is evident that modern farm technology and socio-economic development are entwined in a mutually supportive relationship. The two together form an interesting cycle where each is a means of achieving the other. The message is unambiguous. Advance farm technology undoubtedly leads to socio-economic well-being, and which in turn sets farm technology on wheel towards a greater well-being. And, thus the cycle remains in motion.

Aim and Objectives

The basic aim and objectives of this paper are varied and many which could be summarised as follows:

(i) to find out the socio-economic background of the rural people and its association with other variables;
(ii) to determine the extent up to which the cultivators have adopted new farm technology;
(iii) to determine, analyse and map socio-economic implications of technological change in agriculture;
(iv) to find out differential level of technological change and its socio-economic implications in a hilly and plain area.

Hypotheses

In order to fulfill the above aim and objectives of the study, the following hypotheses have been proposed:

1. Adopters enjoy higher level of education than non-adopters;
2. Adopters have small size of family than non-adopters;
3. Adopters are more active and involved in politics than non-adopters;
4. Adopters are comparatively more mobile than non-adopters;
5. Adopters have more varied occupations than non-adopters;
6. Per capita income of adopters is more-than non-adopters;
7. Adopters enjoy better housing facilities than non-adopters;
8. Adopters are more exposed to mass media than non-adopters.

Researchs Design and Sources of Data

The study is essentially based on primary data collected from two distinct regions of northern India (Rathore, 1984, and Quaiser, 1984). Five villages have been selected from one of the predominantly agricultural state (Uttar Pradesh), while six villages have been taken from the Himalayan State (Jammu and Kashmir). The selection of the villages is done on the basis of stratified purposive sampling. Three villages from each region were selected which were considered to be most advanced in terms of agricultural technology, and they are called as progressive villages. Similarly, three villages from Jammu and Kashmir and two villages from U.P. were selected which were less advanced in agricultural technology, and they are called non-progressive villages. In fact, these villages belong to Etawah district of U.P. and Baramula district of Jammu and Kashmir. However, altogether 300 respondents selecting 150 from each district have been studied. The respondents have been interviewed according to seven exhaustive survey schedules including geographical setting of the village, personal households, characteristics of the respondents, adoption of agricultural innovations, mass-media exposure, social and economic implications of technological change in agriculture. Eight socio-economic indicators, viz. education, size of family, political awareness and involvement, mobility, occupational structure, per capita income, housing condition and media exposure, have been taken in this study.

Data Processing and Analysis

After careful editing, the answers to the questions were classified into meaningful categories so as to bring out their essential pattern. For the purpose of comparability and quantification, the qualitative informations gathered during field survey are allotted scores according to a well-thought of scoring scheme. After completing the task of scoring to all the respondents in all the included parameters in the progressive and non-progressive villages, the average score of education, political awareness and involvement, mobility, occupation, income, housing and media exposure of respondents were worked out. Although tables themselves speakout the socio-economic implications of technological change in agriculture, analysis of variance is also done for testing the significance of associations between two variables.

Socio-economic Implications

The present study conducted in two distinct regions of northern India comes up with several interesting results : some of them are entirely new, while a few are in conformity with the existing ones in the field. It has broadly been found that the level of technological change in agriculture is very much associated with socio-economic level of respondents. There is considerable inter- and intra-regional disparity in the level of innovativeness among the respondents. Generally the respondents of progressive villages are more receptive to new farm technology leading to their better prosperity as compared to the respondents of non-progressive villages. Similarly, the respondents of the U.P. villages are, by and large, more innovative as compared to their counterparts in the village of J&K. Same is true with regard to the socio-economic implications.

However, of the eight hypotheses proposed, six stood the test of empirical analysis and were approved with a fairly high degree of significance. Of the remaining two one is disapproved, while the verdict in case of other is turned out to be tentative. The innovative farmers were found having a clear edge over their traditional counterparts in terms of education, political involvement, mobility, occupation, housing, income and media exposure. However, a brief account of each of them is worthwhile here.

Level of Education

Education is the aggregate of all the processes by means of which a person develops abilities, attitude and other forms of behaviour of positive value in a society he lives (Mohammed and Majeed, 1978). It is essential to cultivate mind so that new ideas once perceived may germinate and strike root. It brings about drastic change in human thinking which plays a catalyst role in production process through organisational and managerial ability. It plays a vital and pivotal role in rural transformation. It is a significant correlate of adoption. It facilitates the adoption of agricultural innovations, which in turn increases agricultural productivity and consequently the economic prosperity.

Formal and informal education has been taken into account for the assessment of the educational level. About 22 per cent respondents are illiterate and 13 per cent have informal education. Among the literate respondents most of them are high school or intermediate pass. There are very few number of graduates and almost all of them belong to the progressive group of respondents.

Since the respondents were having differential levels of educational attainment which is not directly comparable, a scoring scheme has been used to workout the educational level of each respondent. On the basis of individual educational scores, the respondents of both the groups have been put into three categories viz. high, moderate and low, and category-wise average educational score is worked out (Table 12.1).

A perusal of Table 12.1 reveals that the adoption behaviour of cultivators considerably affects their level of education. The adopter respondents of both the regions have generally higher educational scores as compared to non-adopters in all the three categories. By and large, the number of respondents in highly educated category among the adopters and non-adopters is less in both the regions, but in medium and poor category it does not follow any specific pattern. In case of average educational scores, it keeps on decreasing from high to low categories in both the regions with the exception of low category of non-adopter respondents of J&K. when it is more than moderately educated category. There is a considerable difference in the average adoption score of respondents of progressive and non-progressive villages of both the regions, but it is higher in U.P. villages as compared to J&K. The gap becomes even wider when we look at the average educational score of the highly innovative respondents and least innovative ones. The pattern projected by the preceding analysis clearly establishes the relationship between the degree of involvement and level of education. To effect the quantitative expression of exact degree of relationship analysis of variance is used and the results thus arrived, are given in Table 12.2 which shows a much higher value of 'F' (57.2374) as against the table value of F (4.7402) at 0.99 level of significance.

TABLE 12.1: Innovativeness and educational level

Category	Plain Region			Mountainous Region			Total
	High	Medium	Low	High	Medium	Low	
Adopters:							
No. of Resp.	20	46	24	8	20	47	165
Av. Ed. Score	3.25	2.93	1.37	2.96	2.53	1.03	
Non-Adopters:							
No. of Resp.	11	23	26	2	12	61	135
Av. Ed. Score	2.61	2.11	1.09	2.01	0.97	1.02	
Total	31	69	50	10	32	108	300

TABLE 12.2: Analysis of variance of education

Sources	Degree of Freedom	Sum of Squares	Means Sum of Squares	'F'
Between Classes	2	130.6502	65.3251	57.2374
Witnin Classes	297	338.9661	1.1413	
Total	299	469.5270		

A review of the preceding analysis, therefore, strongly confirms the hypothesis we began with.

Political Awareness and Involvement

To-day the human destiny expresses itself in political terms. The politics is so deeply interwoven in the fabrics of our daily life that no facet of it can remain in political isolation and agriculture is no exception to it. The political behaviour of the Indian masses can no more be taken for granted but they have to be wooed, conversed and cajoled. They have gradually learnt the art of claiming their share of the national cake. Having realised the significance link between the political awakening and involvement and socio-economic development, the peasants exert their price before election to extract promises and assurances. Consequently the highly politically conscious regions have cornered the lion's share of the social amenities, while in the other region with politically ignorant people are backward and stagnating.

Generally the economically well of peasants are more aware and involved in political affairs from national to local level. In the present study, therefore, it is hypothesised "Progressive farmers are politically more aware and involved than non-progressive farmers". For the assessment of political awareness and involvement of the farmers, a number of questions relating from international to local politics were put to them at the time of interview. Since these indicators were of varied nature and character, weighted scores were allotted in order to assess the level of political awareness and involvement of each respondent. On the basis of their score, the respondents of both the regions have been put to three categories viz. high, medium and low and their average score of political awareness and involvement is worked out (Table 12.3).

A cursory glance at the Table 12.3 reveals many interesting results. However, the level of political consciousness is satisfactory in the region and is, by and large, in accordance to our proposed hypothesis. The more innovative farmers are found to be more aware of political and social scene and also have greater political involvement. The increase in innovativeness gives rise to the degree of political involvement as well. Therefore, the degree of knowledge and contact with officials and non-officials has a positive bearing on the adoption behaviour of the farmers. There is a inter- and intra-regional disparity in this case as well. It has been found that the respondents of the progressive villages of both the regions are politically more involved and aware of as compared to the respondents of non-progressive villages. Similarly on the whole, the farmers of J&K. have more political consciousness and involvement as compared to the farmers of U.P. It is, perhaps, due to special political status to J&K. and its border with Pakistan.

TABLE 12.3: **Innovativeness and political awareness and development**

Category	Plain Region			Mountainous Region			Total
	High	Medium	Low	High	Medium	Low	
Adopters:							
No. of Resp.	20	36	33	23	21	32	165
Av. Pol. Score	2.81	2.10	1.82	3.01	2.83	1.97	
Non-Adopters:							
No. of Resp.	9	23	28	12	25	38	135
Av. Pol. Score	2.03	1.63	1.15	1.79	1.89	2.01	
Total	29	59	61	35	46	70	300

In order to test this relationship further, the data was subjected to statistical analysis, which strongly establishes a significant positive relationship between innovativeness and political awareness and involvement. The calculated 'F' value is much higher than table value of 'F' and is significant at 1% level.

TABLE 12.4: **Analysis of variance of political awareness and development**

Sources	Degree of Freedom	Sum of Squares	Means Sum of Squares	F
Between Classes	2	65.2404	32.6204	77.6327
Within Classes	297	129.7994	0.4202	
Total	299	195.0398		

The above analysis carried out at various levels and different angels proves our postulation that advanced agricultural technology and political awareness and involvement are strongly and positively related.

Housing Condition

Of all the facets of human environment few influence man's life more than dwellings in which he seeks shelter, security, comfort and dignity. The house in an individual's identity symbol reflecting his social prestige and status. A suitable house is one of three basic needs of human life without which one cannot conceive of a life worthliving. It influences one's physical health and mental efficiency, therefore, future skill and productivity which ultimately determines individual's socio-economic status. Since agricultural advancement has brought economic prosperity, farmers have invested quite a lot in improving their housing condition.

Housing is an universally complex commodity and it is difficult to compare them in totality. Even if they look the same, they are infinitely varied in character and quality and whatever single measure is adopted, none by itself appears to reflect all variations attributed to house size, location, architectural type etc. It is even more complex in rural areas. With this view various components of housing such as storage, number of rooms, courtyard, areal coverage, building material, arrangements for livestocks etc. have been taken in this study. Since the responses of the respondents were varied, various attributes of housing were converted into scores in order to get quantitative idea about the housing condition of the individuals. On the basis of the housing scores of the individuals, the adopter and non-adopter respondents of both the regions have been divided into three categories and the average housing score of each category is worked out (Table 12.5).

A cursory glance of Table 12.5 indicates that on the whole the progressive respondents of both the regions have better housing conditions as compared to non-progressive respondents.

There are a lot of variations in the housing conditions of the respondents. The adopter respondents of the plain region have, by and large, better housing conditions with an average score of 4.01, while the non-adopters have generally poor housing condition, with an average housing score of 2.87. The housing condition of the respondents of the mountainous region is inferior as compared to the respondents of the plain region. Moreover, the number of respondents having better housing condition in mountainous region is also small and majority of them have inferior houses.

TABLE 12.5: Innovativeness and housing conditions

Category	Plain Region			Mountainous Region			Total
	High	Medium	Low	High	Medium	Low	
Adopters:							
No. of Resp.	29	34	27	8	32	35	165
Av. Housing Score	4.21	3.21	3.42	4.00	3.72	2.91	
Non-Adopters:							
No. of Resp.	11	19	30	4	34	37	135
Av. Housing Score	3.01	2.56	1.79	3.14	2.65	2.09	
Total	40	53	57	12	66	72	300

TABLE 12.6: Analysis of variance of housing condition

Sources	Degree of Freedom	Sum of Squares	Mean Sum of Squares	F
Between Classes	2	86.5031	43.2511	27.2130
Within Classes	297	472.0221	1.5893	
Total	299	558.5253		

In order to test this tabular result statistically an analysis of variance has been used and the results are given in Table 12.6. The calculated 'F' value is much higher than the table value. Hence, our proposed hypothesis is strongly and distinctively established. Thus the adoption of improved farm practice undoubtedly leads to an improvement and enlargement of housing facilities. Progressive farmers are normally status conscious, house being a status symbol often gets top position in their list of priorities.

Level of Mobility

It is a process by which an individual is exposed to the world outside his community in person or physically. Mobility usually symbolises progressiveness. Immobility leads to stagnation while mobility paves the path of progress. A mobile farmer gets exposed to a wide range of human endeavour. Repeated exposure to new ideas and increasing familiarity with latest technology speeds up the process of adoption by reducing the time of gestation. It provides impetus towards greater socio-economic consciousness which subsequently leads to unending process of experimentation and action. Recognising mobility as an unfailing indicator of socio-economic development it is hypothesised that progressive farmers are comparatively more mobile than traditional farmers.

For the assessment of level of mobility certain well defined indicators were taken, and frequency of visit and mode of transport of respondent was noted. As per scoring scheme the total scores of each respondent was worked out and the adopter and non-adopter respondents of both the regions were grouped into three categories and the average score of each category was calculated (Table 12.7).

TABLE 12.7: Innovativeness and level of mobility

Category	Plain Region			Mountainous Region			Total
	High	Medium	Low	High	Medium	Low	
Adopters:							
No. of Resp.	33	36	21	27	27	21	165
Av. Mobility Score	4.82	5.53	4.02	4.63	5.49	4.21	
Non-Adopters:							
No. of Resp.	10	21	29	22	25	28	135
Av. Mobility Score	4.16	3.79	2.88	3.15	2.67	3.37	
Total	43	57	50	49	52	49	300

The overall picture which emerges from Table 12.7 is that there is high correlation between level of mobility and innovativeness of the respondents. By and large, the innovative respondents have higher mobility score than non-adopters. Moreover, the highly mobile respondents of both the regions have higher mobility scores and it goes on decreasing from high to low mobile category. The other interesting result is that the respondents of plain region are more mobile as compared to mountainous region which may be contributed to developed means of transport due to flat topography in plain area.

In order to test this relationship statistically, analysis of variance has been used and the result is given in Table 12.8 which is self-explanatory. The calculated value of 'F' is much higher than the table value which shows that the proposed null hypothesis is valid and there is a strong positive relationship between degree of mobility and level of innovativeness.

TABLE 12.8: Analysis of variance of mobility

Sources	Degree of Freedom	Sum of Squares	Mean Sum of Squares	'F'
Between Classes	2	434.2273	217.1136	87.2537
Within Classes	297	739.0469	2.4883	
Total	299	1173.2742		

Per Capita Income

Ever since Independence, the government has been trying to raise the living standard of rural folk to a healthy level of prosperity and well-being, but it began to bear fruit in mid-sixties, and the Green Revolution became manifest with high production leading to relative prosperity to the farmers. It gave rise to gradual attainment of higher standard of living. However, per capita income calculated at current and constant prices is a better indicator to measure the extent of prosperity. It has been, therefore, proposed that adopter respondents have higher per capita income as compared to non-adopter respondents. The per capita income has been calculated by taking into account the earning of the family from all sources and divided by number of its members. On the basis of per capita income, the adopter and non-adopter respondents of both the regions have been divided into three categories and the average income of each category is worked out. Table 12.9 which shows that the adopter respondents register a high per capita income than their counterpart. There are 28 adopter respondents in the higher income group whereas only 4 non-adopter respondents are in this category. Almost similar pattern is in mountainous region where 18 adopter respondents are in higher income group as against only 4 in non-adopter category.

For ascertaining this relationship statistically, the data has been subjected to analysis of variance and the result is shown in Table 12.10 which shows that the calculated value of 'F' is higher than the table value. Hence, the proposed null hypothesis is approved.

TABLE 12.9: Innovativeness and per capita income

Category	Plain Region			Mountainous Region			Total
	High	Medium	Low	High	Medium	Low	
Adopters:							
No. of Resp.	28	30	24	18	20	37	165
Av. Per Capita income	312	266	281	297	281	243	
Non-Adopters:							
No. of Resp.	4	30	26	4	18	53	135
Av. Per Capita income	297	254	186	285	272	281	
Total	32	68	50	22	38	90	300

TABLE 12.10: Analysis of variance of per capita income

Sources	Degree of Freedom	Sum of Squares	Mean Sum of Squares	'F'
Between Classes	2	999072.430	499536.215	18.271
Within Classes	297	8120094.600	27340.383	
Total	299	9119167.030		

Level of Exposure to Mass Media

Among the various means of communication, mass communication plays a significant role in the social change as it reaches the audience more easily, and affects knowledge, attitude, opinion and behaviour of individuals (Lazarsfield, 1951). Exposure to mass media is an irrefutable indicator of socio-economic development. In order to investigate into the nature of underlying correspondence between the adoption behaviour of the farmers and the degree of their exposure to the household agent of mass communication a null hypothesis viz. "adopter respondents are more exposed to mass media than non-adopters" is proposed.

For the assessment of exposure to mass media–radio/transistor, magazine, newspaper, motion picture, fair and exhibitions and televisions have been taken and to facilitate comparability of data and its subsequent quantification, weighted scores were allotted to different components of mass media. The total score of each respondent is worked out. The adopters and non-adopters of both regions have been classified into three categories and the average exposure score of each category has finally been worked out (Table 12.11), which is a self-explanatory. A substantial difference was perceived between the exposure level of adopter and non-adopter respondents. There is considerable difference in the level of exposure of respondents of plain and mountainous regions. The adopter respondents of former are more in high level exposure category, while in the latter there is only one respondent in this category and 47 respondents in the lower category. The pattern of level of exposure of non-adopters is just reverse in plain region viz. lesser number of respondents are highly exposed, while large number is in least exposed category. In the mountainous region it is similar to adopter group of respondents.

TABLE 12.11: Innovativeness and exposure to mass media

Category	Plain Region			Mountainous Region			Total
	High	Medium	Low	High	Medium	Low	
Adopters:							
No. of Resp.	39	32	19	1	27	47	165
Av. Expo. Score	4.67	4.73	3.66	3.12	3.49	3.12	
Non-Adopters:							
No. of Resp.	12	18	30	1	20	54	135
Av. Exp. Score	3.82	2.17	3.01	2.11	3.25	2.46	
Total	51	50	49	2	47	101	300

In order to get exact degree of relationship, the analysis of variance has been used and the result is given in Table 12.12.

TABLE 12.12: Analysis of variance of exposure of mass media

Source	Degree of Freedom	Sum of Squares	Mean Sum of Squares	'F'
Between Classes	2	176.2493	88.1246	68.1272
Within Classes	297	389.9127	1.3128	
Total	299	566.1630		

The calculated 'F' value establishes the co-existence of the two variables as the result is significant at 1 per cent level. Since the calculated value of 'F' is higher than its table value, the proposed null hypothesis is proved.

Occupational Structure

Owing to its vulnerability to vagaries of weather conditions and technological advancement, agriculture has no longer remained the safe occupation of farmers. Not only to secure his livelihood but also to become economically more prosperous, the farmers branch out to diverse other occupations such as business, rural craft, govt. and semi-govt. jobs and so on. Some are their traditional occupations, while some are entirely new. The comparative economic prosperity due to induction of technology in agriculture has resulted in spread of education and setting up business ventures. The educated rural youth has gone to urban centres to look for a job in government or private concern. Besides this, all the incentives and subsidies meant for small and marginal farmers have been, by and large, gobbled up by rich farmers who have political links and familiarity with government officials. And this fund has been misappropriated and diverted towards setting up new business.

However, by and large, it is observed that the innovative farmers have more varied occupation. And, therefore, a hypothesis — "Progressive farmers have varied occupational structure than their counterpart" has been putforth.

For the measurement of variety in occupations, several questions related to agriculture, dairy, fishing, service, craft etc. were put to respondents at the time of interview. A cursory glance at the answers indicates that the farmers have adopted varied occupations, but the occupational structure of innovative respondents is more varied and complex. Since the occupations are of varied nature and are not comparable, a scoring scheme evolved in proportion to income generated by each one of them, was used and the total score of all the respondents have been worked out and finally on the basis of it the progressive and non-progressive respondents of both the regions have been put into three categories each viz. highly varied, moderately varied, and less varied, and the average occupational score of each category is worked out (Table 12.13). The analysis reveals that innovative respondents have more average adoption score than their counterpart. The other interesting result is that the respondents of J&K have more varied occupations and have higher average scores as compared to the respondents of the plain region. The relationship in the plain region is not very strong and clear-cut.

TABLE 12.13: **Innovativeness and occupational structure**

Category	Plain Region			Mountainous Region			Total
	High	Medium	Low	High	Medium	Low	
Adopters:							
No. of Resp.	29	28	33	7	14	54	165
Av. Score	3.50	3.22	2.79	3.49	3.01	2.17	
Non-Adopters:							
No. of Resp.	15	14	30	8	9	59	135
Av. Score	2.83	3.01	1.73	2.72	1.96	1.86	
Total	44	42	63	15	23	123	300

However, in order to probe further into it analysis of variance has been applied and the results are given in Table 12.14.

TABLE 12.14: **Analysis of variance of occupational structure**

Sources	Degree of Freedom	Sum of Squares	Mean Sum of Squares	'F'
Between Classes	2	243.2650	121.6325	4.2130
Within Classes	297	7574.5682	28.8706	
Total	299	7817.8332		

At a cursory glance of Table 12.4 shows that the calculated value of 'F' is slightly higher than its table value. And it proves our hypothesis viz. progressive farmers have more varied occupation than non-progressive farmers.

Size of Family

It is usually believed that spread of education tends to act as a brake upon population upsurge. A positive correlation between innovativeness and education being hypothesis of this investigation, the inclination of the innovative farmers towards a planned family appeared but a natural extension of the same. Moreover, the innovative farmer's openness to urban influences makes him aspire for a higher standard of living which is not compatable with large families. On the other hand, traditional farmers are incurable fatalists taking children as gift of God. Family planning to them amounts to interference with the divine scheme. Family planning for them is just as another innovation which should be discouraged. With this view a hypothesis "adopter respondents have small sized families than their counterpart" was formulated to least.

The results of investigation proved eye opener in more than one way. Certain progressive families were found abnormally large, while a number of laggards have reasonably planned families. However, on the basis of size of families adopter and non-adopter respondents of both the regions have been put into three categories i.e. large, medium and small and their average size of family is worked out (Table 12.15).

TABLE 12.15: **Innovativeness and size of family**

Category	Plain Region			Mountainous Region			Total
	High	Medium	Low	High	Medium	Low	
Adopters:							
No. of Resp.	23	42	25	27	37	11	165
Av. family size	11.20	9.33	12.65	12.95	13.01	10.71	
Non-Adopters:							
No. of Resp.	23	26	11	26	38	11	135
Av. family size	12.30	3.01	11.97	11.20	12.10	13.52	
Total	46	68	36	53	75	22	300

This table gives unexpected picture. There is no definite relationship between the innovativeness and size of families of respondents of both the regions or even at aggregate level. A mixed and complicated pattern has emerged which shows that two variables stand independent of each other.

In order to reaffirm this viewpoint, analysis of variance is applied for a finer analysis and the result is given in Table 12.16. The calculated value of 'F' is less than the table value. Thus our proposed null hypothesis is totally rejected.

TABLE 12.16: **Analysis of variance of size of family**

Sources	Degree of Freedom	Sum of Squares	Mean Sum of Squares	'F'
Between Classes	2	328.5143	164.2571	1.1915
Within Classes	297	39942.0440	137.8520	
Total	299	40270.5583		

Conclusion

Leaving exceptions apart, the overall picture which emerges from the present study is that this improved farm technology acts as unfailing engine to accelerate the pace of socio-economic development. Several studies conducted so far also reveal that the trend and process of modern farm technology is sped up with a number of socio-economic factors such as education, income, training, size of holding, tenancy system, timely and controlled water supply, adequate credit provision, incentives and subsidies. The present paper approaches the problem from the reverse side. It proves that the induction of modern farm technology brightens up socio-economic picture. It is, therefore, evident that modern farm technology and socio-economic development are entwined in a mutually supportive relationship. The two together form an interesting cycles where each is a means of achieving the other. The message is crystal clear, advanced farm technology undoubtedly leads to socio-economic well-being and which in turn sets farm technology on wheels towards a greater well-being, and so the cycle remains in motion.

Thus the devil of rural backwardness could be banquished by adopting the strategy consisting of a two pronged attack i.e., large scale induction of modern agricultural technology, and simultaneous efforts towards creation of a socio-economic climate congenial to it. The former is possible when an efficient agricultural infrastructure is developed, while the later could be created by raising the level of education, maximising the exposure of farmers to mass media, and revitalising the near defunct extension services.

REFERENCES

Casette, E., 1981 : 'Correlates and Determinants of Agricultural Productivity Growth : A test of some Hypothesis', in N.Mohammad (Eds.), *Perspectives in Agricultural Geography*, Vol. IV, Concept Pub. Co., Delhi.

Dandekar, V. M., 1964 : Regional Variations in Agricultural Development, *Indian Journal of Agricultural Economic*, Vol. 19.

Dharma, N., 1977 : Growth of Productivity in Indian Agriculture, *Indian Journal of Agricultural Economics*, Vol. 32.

Marsh, C. P. and Coleman, A. L., 1966 : 'Group Influences and Agricultural Innovations' : Some Tentative Findings and Hypotheses, *American Journal of Sociology*.

Mohammad, N. and Majeed, A., 1978 Impact of Socio-economic Factors on Diffusion of Agricultural Innovations', in A. Mohammad (Eds.), *Dynamics of Agricultural Development*, Concept Pub. Co., Delhi.

Qaisar, S. H., 1984 : *Socio-economic Implications of Technological Changes in Agriculture* (Unpublished Dissertation, Delhi University)

Rao, C. H. H., 1975 : *Technological Change and Distribution of Gains in Indian Agriculture*, Mac Millan, Delhi.

Rathore, B. K., 1984 : *Socio-economic Implications of Technological Changes in Agriculture* (unpublished Dissertation, Delhi University).

Shafi, M., 1984 : *Agricultural Productivity and Regional Imbalances*, Concept Pub. Co., Delhi.

13

IMPACT OF GREEN REVOLUTION ON SUBSISTENCE FARMING IN WESTERN RAJASTHAN

SALAHUDDIN QURESHI

THE STUDY seeks to identify shadow effects of Green Revolution on subsistence Farming. It brings out that Green Revolution inflicts severe economic stresses on the survival mechanism of subsistence farming. The problem has been screened in Ganganagar district of arid western Rajasthan as a case study. Green Revolution involves the evolution of new strains of wheat in northwestern part of India and the impact of cultivating new strains. The evolution of high yielding varieties is the outcome of research work and plant breeding achievements. Green Revolution has enabled spectacular increase in the yields and double cropping pattern in many areas.

It has been emphasised that the infrastructural and fertiliser costs of production may be considerably large and the increased yield do not necessarily produce sufficient revenue to recover the costs when market prices stabilise or fall with increased supply. However, such a contention may hold good only in countries where large surpluses have been obtained. In India, where scarcity market conditions still prevail such a situation cannot be accepted in principle without evaluating it practically. Priority in development should be given to the area which is facing more intense competition of survival and where subsistence farmers suffer progressive economic losses.

Green Revolution on its way to India not only encountered the readily reckoned technical and economic problems but also faced very complicated and stubborn social and behavioural problems. Social inhibitions are more difficult to eradicate. Although the attempts of conveyance and implementation of Green Revolution were thawarted by many sociological aspects of our society, Green Revolution, however, succeeded in establishing its footholds in the Indian agriculture. Increased production is an economic aspect and is easier to achieve through improved technology. However, judicious distribution is a social aspect and is much more difficult to attain.

Green Revolution in its earlier stages concealed the inherent signs of its opposition to subsistence farming. At the onset of Green Revolution much of the agricultural lands were subsistence and were helped by the revolution. It was the period when agriculture was not as much market-oriented. The survival mechanism of subsistence farming was weakened when the suitable ratio of production under commercial versus subsistence farming was considerably tilted towards commercial farming. Thus commercial farmers began to create market- and price environment unfavourable to subsistence farmers. It is necessary to clarify the point that although subsistence farming is not a market farming but it cannot escape the market situation/costs in the procurement of various inputs.

Irrigated farmers have not to face the intensity of weather vagaries on comparative scales with subsistence farmers. They possess an initial and outright advantage of ensured water supply. The subsistence farmers have to procure the input ingredients at the market rates but their limited and uncertain output does not match the market mechanism, rendering their agricultural system vulnerable to economic decay. The government subsidies are inadequate and untimely.

The Resource Aid Schemes in Green Revolution were comparatively more tangible for the already established large farmers. As a result of it they capitalised over the incentives. On the other hand, the compensatory relief schemes for subsistence farmers such as Small Farmers Development Agency (SFDA) and Marginal Farmers and Agricultural Labour Agency could not fetch relief due to the inbuilt flaws in the mode of relief. Landuse change in Ganganagar district has shown market dominance. The effect of Green Revolution in the district has resulted a decrease in the proportion of cultivators. While an increase in the agricultural labour has been registered. It means that the marginal subsistence farmers are increasingly becoming the landless agricultural labour. The marginal farmers find the selling out of their economically unfeasible land, because of input deficiency, to big farmers more profitable than to retain it with their meagre inputs. Thus the marginal lands have been rehabilitated by Green Revolution but the marginal farmers have been abandoned, and are sliding down on the economic ladder. Such marginal farmers become landless agricultural labour as well as the urban migrants on competition with commercial irrigation farming the subsistence dry land agriculture is becoming uneconomic proposition in Ganganagar district.

The fact that the increased hectarage and yield in commercial variety of rice cultivation in the desert district of Ganganagar where rice is not a staple food of the people is an indicator of market dominance and not of social welfare (Fig. 13.1). Moreover, the cultivation of water intensive crop like rice is not ecologically suitable in arid district of Ganganagar. Although there is no direct competition of prices between the Green Revolution crops and the subsistence crops in Ganganagar district as the two types of crops are of different nature, yet the surplus crop produce of commercial farmer by virtue of induced inflationary conditions inflicts an adverse effect on the cost-benefit ratio of subsistence farming. The competitive low market prices of subsistence *bajra farmer* against commercial *wheat farmer* push the former under economic distress. The 1982 price environment of various crops in different tehsils of Ganganagar district is given below.

In Suratgarh tehsil the bajra (pearl millet) market price was Rs. 140 quintal against the wheat price of Rs. 180 quintal. Similarly, in Nohar and Bhadra tehsils, the two largest bajra growing tehsils, the bajra price was only Rs. 100 quintal against that of Rs. 180 quintal for wheat. In Tibi and Hanumangarh tehsils also bajra price was Rs. 120 quintal against wheat price of Rs. 190 quintal and Rs. 175 quintal. The price structure of bajra is governed by two factors.

(a) Hectare under bajra, and
(b) Price effect of higher percentage of irrigation and commercialisation in Green Revolution crops.

Thus, it could be concluded that Green Revolution along with its production surpluses and resultant prosperity to commercial farmers has made survival of subsistence farmers more difficult than before. The economic condition of the subsistence dryland farmers in Ganganagar tehsil is miserable because subsistence farmers have only

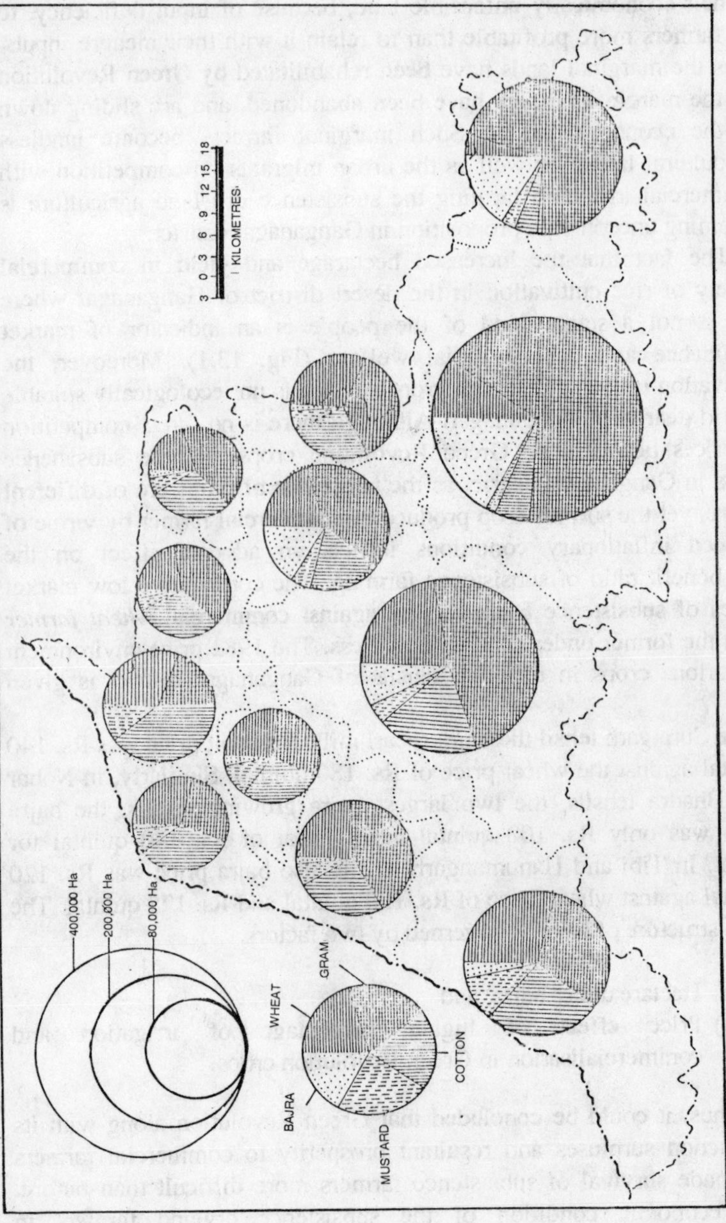

FIG. 13.1: Ganganagar District Rajasthan: area under major crops.

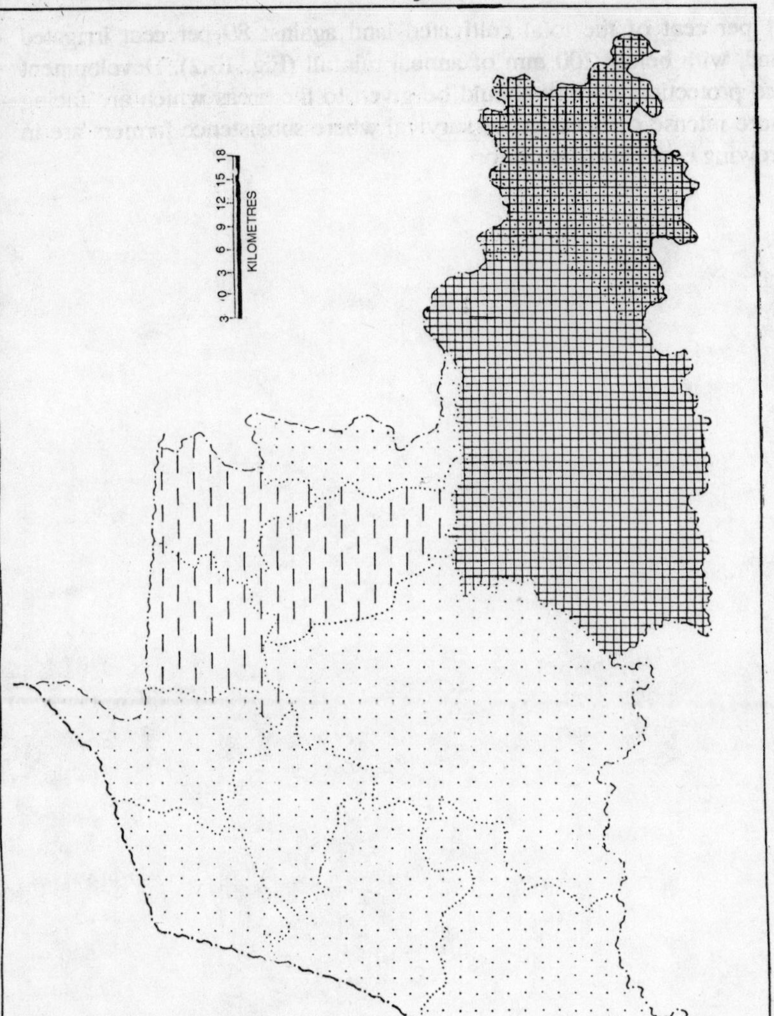

FIG. 13.2: Ganganagar District Rajasthan: average annual rainfall.

11 per cent of the total cultivated land against 89 per cent irrigated land, with below 200 mm of annual rainfall (Fig. 13.2). Development and protection priority should be given to the areas which are facing more intense competition of survival where subsistence farmers are in growing economic discomfort.

14

CENTRAL PLACE FUNCTION AND GROWTH CENTRE POTENTIAL OF SMALLER URBAN PLACES IN RURAL AREAS

RICHARD E. PRESTON

AT LEAST two sets of problems arise in the implementation of regional development strategies based on growth centres. One group involves the identification and selection of growth centres, while a second group is related to the nature of interaction between growth centres and their surrounding areas.[1] The general goal of this study is to consider ways how ideas from the classical Central Place Theory might contribute to solutions to both sets of problems.[2] The specific purpose, however, is to suggest that an answer to one of the Thornier problems facing decision-makers responsible for selecting growth centres may be found in the Central Place Theory. Accordingly, an approach to the problem of *determining the growth centre prospects of smaller urban places* that is quite different from present procedures is considered and demonstrated.[3] Because the emphasis of this collection of papers is on agricultural geography, attention is given to the role of growth centres in rural development.

Growth centres are defined here as *locations* that offer potential for future economic and social development, that is, places where conditions necessary for expanding employment opportunity either exist or can be efficiently created. Such growth centres are not only expected to improve conditions for local residents, they are expected to attract commuters and migrants from surrounding areas of labour surplus and to stimulate employment growth in some of those areas.

Smaller places are arbitrarily defined as those with corporate city populations of less than 50,000.[4] It is not suggested that all growth centres should be chosen from existing smaller places or that smaller places have been ignored in growth centre programmes. The case for the designation of at least some larger places as growth centres has been argued convincingly, and whether or not smaller centres are considered depends on the objectives of particular programmes.[5] Rather, it is suggested that in schemes where smaller centres have been included, little success has been experienced in the selection of particular places for favoured treatment. For example, the Economic Development Administration, U.S. Department of Commerce, attempted to apply a growth centre strategy on a national scale. It was evaluated as unsuccessful because development funds were too widely and thinly dispersed over too many small places (Hansen, 1975).

The apparent failure of smaller places as growth centres suggests either that there is a limited role for such places in development programmes or that there is an absence of reasonably objective and politically acceptable procedures for estimating and comparing their growth centre potential. It is argued here that there is a role for smaller growth centres in development programmes, particularly in those focusing on rural areas, and that a significant barrier to the effective use of smaller growth centres is just such a lack of credible growth centre potential evaluation procedures. Moreover, it appears that without such procedures a virtual Pandora's box is opened whereby every small-town, political and business development group can demonstrate that their community's greatness awaits only a well-timed financial boost from government (Hoover, 1975).

Procedures used to determine growth centre potential appear to inadequately evaluate the prospects of smaller places for at least two reasons. First, indicators that satisfactorily measure the fortunes of larger places may measure those of smaller places less well. For example, of the indicators commonly used to evaluate growth, centre potential, like population size, rate of population growth, job formation in non-primary activities, importance as a focus for migrants, and change in the number of functions or establishments, only rate of population growth does not introduce a bias toward larger places (Hansen, 1971a). A second and related reason is that because evaluation procedures give undue importance to population size, say, to 50,000 or greater as a designation of metropolitanism, the functional importance of at least some smaller places in the organisation of space economies is probably underestimated. These points are developed below.

Pertinence

The growth centre potential of smaller places is an important issue facing regional development programmes for at least three reasons. First, inability to solve the problem which has led to advocation of growth centre selection procedures that possibly overemphasise the importance of population size as a key index of economic potential. The population size argument rests on the belief that: (1) communities with labour shed populations of 250,000 or more and urban cores of 50,000 or more have adequate labour pools, infrastructure and external economies to meet the needs of most new activities; (2) that few communities with less than 50,000 people possess adequate labour, infrastructure, or external economies to attract new firms and that few places of this size can influence the welfare of their surrounding communities; and (3) that places with more than 250,000 inhabitants possess conditions necessary for self-sustaining growth and are less in need of aid. This position urges that efforts at growth centre development be concentrated in places in the 50,000 to 250,000 size range, and that smaller place be ignored unless special conditions are present (Hansen, 1971; Berry, 1979). This practice has resulted in what could be an unjustified condemnation, or ignorance of potential contributions smaller places might make to regional development. Although the likelihood of some further growth is undoubtedly greater in some larger places than in most smaller ones, considerable uncertainty surrounds the views that, first, there is a minimum viable size for growth centres; and, second, that there is a critical threshold beyond which self-sustaining growth is assured (Moseley, 1974).

A second reason why the growth centre potential of smaller places is important is simply because vast inhabited sections of North America lie in the outer reaches of or beyond the daily urban systems of places of 50,000 or larger and larger sections of such areas are experiencing economic difficulty (Morrill, 1973a, 1973b; Simmons, 1976). A third reason is that funds available for regional development programmes are always limited. So, if growth centres are to have the greatest positive impact on problem areas, a limited number of places having substantial development potential must be selected.

The Climate of Research on Growth Centres and the Relationship between Growth Centres and Rural Development

Over the past 20 years the growth centre concept has experienced a cycle of boom and slump (Richardson, 1978). Not long ago growth centres were considered to offer a solution for numerous regional development problems in developed and developing countries alike. As a result of this belief, policies supporting deliberate attempts to channel economic growth into certain locations were either advocated or practised in a number of nations (Moseley, 1974; Kuklinski, 1972). Recently, however, resistance to development strategies employing growth centres has intensified, especially in developing countries (Richardson, 1978; Friedmann and Weaver, 1978). Not only is the concept no longer regarded as a panacea for regional problems, its value as a constructive tool to be used at any point in the regional development process has been questioned. Richardson reviewed the changing fortunes of the growth centre concept and concluded, first, that while development programmes based on growth centres do not offer a universal remedy for regional problems everywhere, growth centres remain a valuable component of national economic development programmes employing a planned urbanisation strategy; and, second, that growth centres are particularly well suited for use in rural development schemes.

Richardson gives reasons for his views (Richardson, 1978). Regarding present widespread dissatisfaction with the growth centre concept, he argues that the apparent failure of growth centres programmes has little to do with the value of the concept as a basis for regional development. He suggests, for example, that: (1) insufficient care has been given to the selection and designation of growth centres, (2) planners have failed to integrate growth centres into over all national spatial development strategies, (3) politicians have unrealistic expectations regarding the length of time required for programmes based on growth centres to bear fruit, and that when results are not achieved quickly, the course of development is often changed unwisely, (4) implementation has been ineffective because of weak pursuit of growth centre strategies by governments initiating them, (5) a tendency to call almost every kind of spatial policy a growth centre approach has concealed weak over all research programmes that provide insufficient bases for implementation of meaningful growth centre strategies, and (6) that there has developed a widespread rejection of the growth centre concept on the basis of an assumed contradiction between urbanisation and rural development. Frequent occurrence of these events has brought about what Richardson suggests is a state of widespread confusion and disarray about what growth centre policies mean (Richardson, 1978).

Regarding relations between urbanisation and rural development, Richardson points out that while there has been an urban bias in development planning, the assumed contradiction between growth centre strategies and rural development does not necessarily exist. In fact he argues that appropriate growth centre strategies may strengthen rural areas rather than divert resources from them (Richardson, 1978). For example, growth centres should anchor key urban systems in areas needing assistance because they can create and maintain both demographic and economic stability throughout existing settlement hierarchies. This is achieved at least partially through reinforcement and upgrading of transportation and communication networks, by firmer administrative control of service functions, and by provision of both full-time and part-time off-farm jobs (Richardson, 1978). Likewise, stable hierarchies may be essential for the delivery of a range of services essential for human resource development. The provision of more and better services is not only pivotal to attempts to raise living standards, the availability of educational, public health, and other social welfare services is possible "interceptor" towns may also influence weather local rural residents stay or migrate to larger cities.

The points made suggest that reasons behind the apparent failure of development programmes based on growth centre strategies have little to do with the actual value of the concept as a tool for regional development, and that the supposed contradiction between rural development and growth centre strategies is more apparent than real.

Procedure

As noted above, it is argued here that an evaluation of the growth centre potential of smaller places can be based on factors quite different from, or at least in addition to, those previously used to evaluate both larger and smaller places. Moreover, ideas from classical central place theory might play a key role in such a task. Central place theory provides a description and explanation of the territorial arrangement of tertiary and institutional activities. Unlike primary and most secondary and quarternary activities, tertiary and institutional functions are present in virtually every settlement. Because of both their ubiquity and the fact that they represent the economic sector where job formation is currently greatest, these activities provide both a basis for community comparison and potential economic development instruments (Hermansen, 1969).

Essential points are, first, that the central place function (the provision of central goods and services in excess of those required by a community's inhabitants) can remain stable or even increase in places where non-central place activities fluctuate in response to both cyclical and secular patterns of demand. Second, central place functions may be able to provide an on-going economic base for places supported by renewable physical resources managed on a sustained-yield basis; for example, for settlements supported by agriculture, forestry, fishing, tourism, and water-related activities like hydro-power, recreation, and navigation. Towns associated with maintenance and renewal of human resources, like military, political, higher educational, retirement and research and development activities might fall in a similar class. Third, some central places will persist and others emerge in the outer reaches of the urban fields of metropolitan areas where they will function as service centres not only for people employed locally but for an increasing population commuting on varied schedules to locations within metropolitan daily labour sheds (Berry, 1970, 1976).[6] A final point is that the ability of the public sector to influence central place development is considerable as the location and improvement of transportation networks and of numerous administrative and institutional functions fall under its jurisdiction (Richardson, 1973b; Berry, 1973).

If the previous consideration has merit, then it seems reasonable to ask, "Why can't a stable or expanding central place function provide a long-term base for the economic persistence or expansion of places with less than 50,000 inhabitants, particularly if employment increments are in reasonable relation to existing infrastructure, labour, and service resources"? As a first step in an examination of this possibility it was undertaken to determine whether such stable or expanding central places exist in an area of officially recognised to be in need of regional economic development, and, if so, to see if they are located in or adjacent to particular areas where growth centres might contribute to regional development.

The Study Area

The study area is the Pacific north-west section of the United States, defined as the States of Washington, Oregon, Idaho, and mountainous counties of Western Montana. The official view of the region's economic health is clear from its inclusion in the area administered either by the Pacific North-west Regional Commission (embracing Washington, Oregon and Idaho) or the old North-west Regional Commission (which includes Western Montana E.D.D., 1978; E.D.A., 1978). These regional commissions are modelled after the Appalachian Regional Commission and their task is to promote regional economic development in its broadest sense (P.W.E.D.A, 1965, 1976; and P.M.R.C., 1977). Commissions identify parts of their territory needing development assistance and provide both short and long-term aid. County economies are continually monitored according to several criteria among which income levels and the state of employment are most important. Counties deemed in economic difficulty are designated initially for short-term aid, and then are grouped into larger Economic Development Districts which receive long-term assistance. In 1978 only three out of the 136 counties in the North-west did not qualify for either long or short-term federal aid.

The North-west is an area with a strong and varied physical geography. Accordingly, population is unevenly distributed, with sparsely inhabited mountains and plateaus occupying a major portion of the land area. Large urban concentrations are few, and a variety of economies are present. Economic development has traditionally been dependent on export-oriented resource-based industries, with logging, lumber, agriculture, food processing and paper products in all States, plus fishing in Oregon and Washington, and mining in Idaho and Montana dominating the region's development. Prior to 1960, with the exception of the aerospace industry represented mainly by the Boeing Company located in Seattle, non-resource oriented manufacturing industries were comparatively small or absent from the region. Although diversification of the economic base began during and after the Second World War with the development of aluminium production based on hydro-electric power, localised chemical industries associated with atomic energy, and a more widespread electronics industry, dependence on resource-based activities remains and is pronounced beyond the commuting zones of major urban centres (P.M.R.C., 1977; Highsmith, 1962, 1973; Preston, 1971a).

Figure 14.1 provides one perspective on the region's urban system. The map shows the central place order of places with 2,500 or more inhabitants in 1960 and the territorial pattern of the urban systems as defined by daily newspapers and banking linkages. This perspective incorporates virtually every place of hamlet size or larger. The area's human geography is dominated by an urban-industrial heartland occupying the Willamette-Puget Sound Lowland, extending from Bellingham, Washington, near the Canadian Border to Eugene, Oregon, in the south. The heartland is in turn dominated by two metropolitan areas, both with populations in excess of one million: the Seattle-Tacoma Complex in the Puget Sound Lowland and Portland-Vancouver in the northern Willamette Valley and adjacent Washington. Metropolitan regions centring on Salt Lake City, Utah, and Great Falls, Montana, exert influence in the North-west although these cities are located outside the region as defined here.

Beyond the heartland, urban places are concentrated in two crescents. One is composed of the agriculturally productive Wenatchee, Ellensburg and Yakima Valleys and the cluster of cities near the junction of the Columbia and Snake Rivers (the Tri-Cities of Kennewick Richland-Pasco and nearby Walla). The other crescent consists of three sections of the Snake River Plain in Southern Idaho. From west to east, the cities of Boise, Twin Falls and Idaho Falls—Pocatello provide foci for each section and for their respective agricultural economies.

A different situation prevails on the sparsely populated and extensive agriculture-oriented central plain and eastern border of the Columbia Basin. This vast area is economically organised around Spokane, the North-west's third largest city (182,000), and supports a dispersed pattern of trade centres like Ephrata, Othello and Colfax in Washington; Pendleton, Milton, Freewater, and Hermiston in Oregon, and Grangeville and Indaho in Moscow.

Most of the remaining urban places have small populations and are either scattered or located in clusters sharing a similar economic base. Places like Brookings, Coos Bay-North Bend, Newport, Astoria, Raymond, Hoquiam-Aberdeen, and Port Anagles occupy small lowlands along the Pacific Coast and dominate limited urban fields. Forest industries, commercial fishing, and some tourism contribute to the economic life of these coastal towns. Throughout the mountains are places like Medford, Roseburg, Grants Pass, and Klamath Falls, all dominating areas in or adjacent to highlands in south-western Oregon,

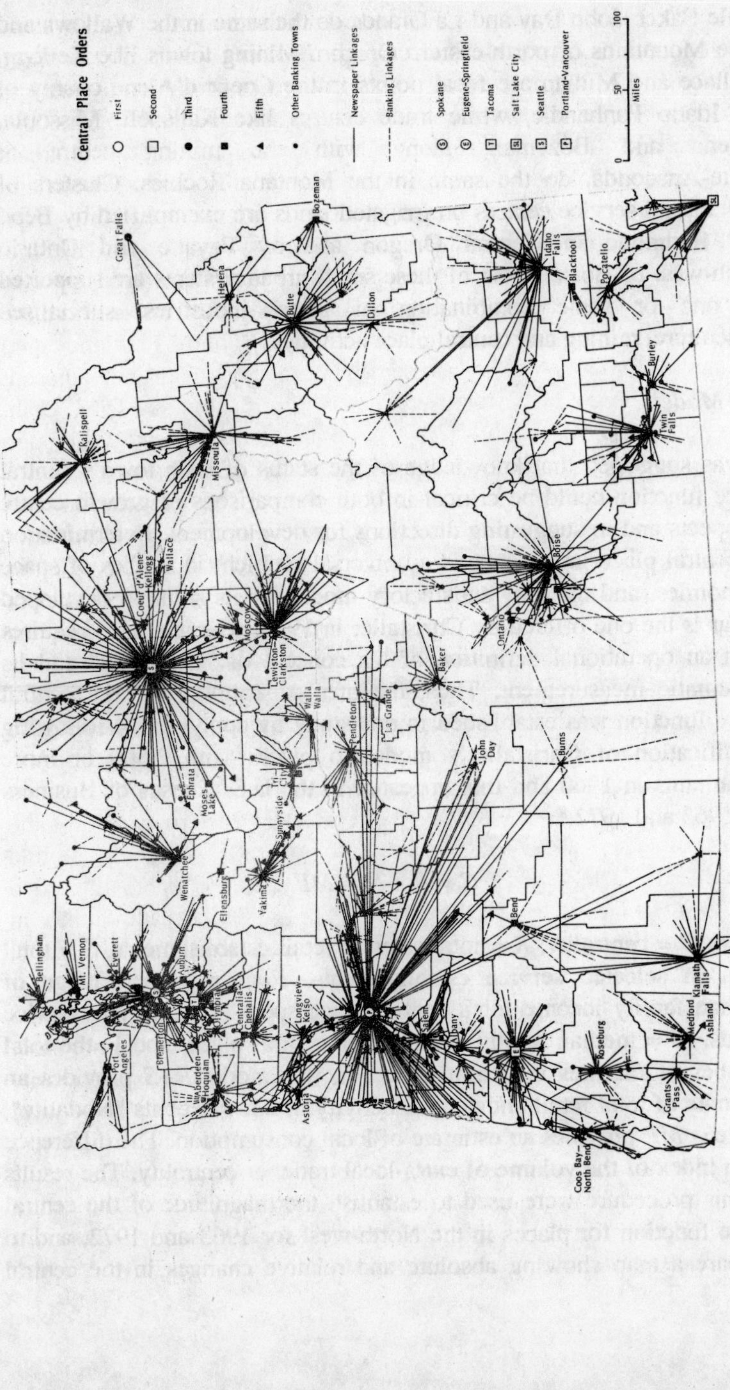

FIG. 14.1: The urban system in the Pacific north-west.

while Baker, John Day and La Grande do the same in the Wallowa and Blue Mountains of north-eastern Oregon. Mining towns like Kellogg, Wallace and Mullan are focal points in the Coeur d'Alene county of the Idaho Panhandle, while trade centres like Kalispell, Missoula, Helena and Bozeman, along with the mining centre of Butte-Anaconda, do the same in the Montana Rockies. Clusters of agricultural service centres on irrigated lands are exemplified by Bend and Redmond in Central Oregon and by Payatte and Ontario north-west of Boise. Most of these small urban systems are supported by one or some combination of forest industries, specialised agriculture, mining and central place activities.

The Model

It was suggested that knowledge of the status of each town's central place function could be critical in both comparisons of growth centre prospects and in suggesting directions for development. Determination of central place importance is a universal problem in studies of space economies, and the only satisfactory model for its solution developed so far is the one offered by Christaller in 1933.[7] His approach requires both an operational definition of his concept of "Centrality" and its systematic measurement. Thus, the areal distribution of the central place function was established in this study by applying the following modification of Christaller's model to places with 2,500 or more inhabitants in 1960 and that appeared in the U.S. Census of Business for 1963 and 1972.[8]

$$C = R + S - \alpha MF$$

where C = centrality, R = total sales in retail establishments, S = total sales in selected service establishments, α=average percentage of median family income for a central place spent on retail and service goods, M = median family income in a central place, and F=the total number of families in central place. In this model $R+S$ provides an estimate of total retail and service activity in a town, or its "Nodality", while αMF provides an estimate of local consumption. The difference is an index of the volume of extra-local trade, or centrality. The results of this procedure were used to establish the magnitude of the central place function for places in the North-west for 1963 and 1972, and to prepare a map showing absolute and relative changes in the central place function between those dates.

Findings

Stability and Change in Central Place Importance

Figure 14.2 shows the distribution of percentage changes in centrality (1962 dollars) by locations between 1963 and 1972. While there were 164 places with 2,500 or more people in 1960 only 128 locations were considered here. Such was the case because some places were absorbed over the decade by expansion of the urbanised areas (continuous built-up areas) of Seattle-Everett, Tacoma, Fortland-Vancouver and Eugene, and because it was considered more realistic to treat adjacent towns like Hoquiam-Aberdeen, Lewiston-Clarkston and Kennewick-Richland-Pasco (Tri-Cities) as single locations.[9] Of the 128 locations considered, 103 had centrality increases. Eighteen places had losses of greater than 10 per cent. Nineteen places experienced changes between -10 and +10 per cent and were considered stable. The largest concentration of percentage increases was between +11 and +69, an interval embracing 77 places. Conclusions here were that most central places in the north-west registered centrality increases between 1963 and 1972, and that such increases were often substantial.

Figure 14.3 shows centrality differences between 1963 and 1972 in 1962 dollars. Centrality differences are shown by circles while percentage changes in centrality are shown by patterns. Outstanding increases in both centrality magnitudes and percentage increases (greater than 70%) were registered by Bellingham and Marysville in Washington; Roseburg, Bend, Ontario and Newport in Oregon; Sand Point, Coeur d'Alene, Rexburg and Nampa in Idaho; and Missoula, Helena and Bozeman in Montana. Of these places, Bellingham, Roseburg, Bend, Missoula, Helena and Bozeman are located either peripheral to or outside of the heartland and are foci for important segments of the region's settlement pattern (Fig. 14.1). Increases in centrality of greater than 39 per cent were recorded by such important non-heartland central places as Medford, Grants Pass and La Grande in Oregon; Yakima and Hoguiam-Aberdeen in Washington; and Moscow and Idaho Falls in Idaho. Other central places occupying critical locations in the region that also registered centrality percentage increases close to 39 per cent (Sauer, 1968) were Kalispell, Montana

(37%), Wenatchee (38%) and Tri-Cities (37%), Washington.[10]

Some conclusions here were, first, that the central place function expanded between 1963 and 1972 for places in all population size groups and that such places were widely distributed. Second, numerous central places with fewer than 50,000 inhabitants that were also focal points for significant settlement systems increased the magnitude of their regional service function by more than 36 per cent. Third, most of the locations experiencing greatest percentage increases were located in agricultural and other resource industry-based areas outside the regional heartland. Fourth, all non-heartland places experiencing either percentage declines greater than 10 per cent or in the -10 to +10 per cent group were small. And, fifth, that proximity of many stable or declining places to locations with expanding centralities suggests that some functions formerly offered by stagnating centres are being offered by growing central places as part of a more competitive functional mix.

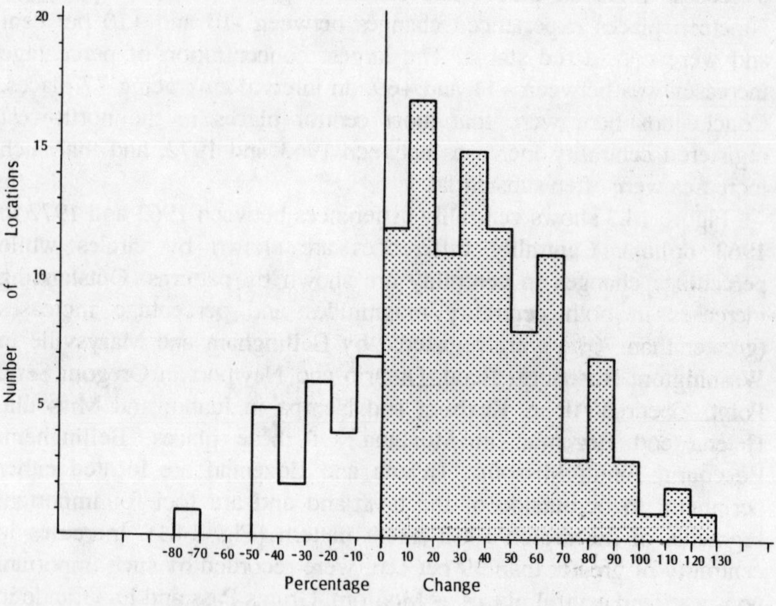

FIG 14.2: Per cent change in centrality by location, 1963—1972.

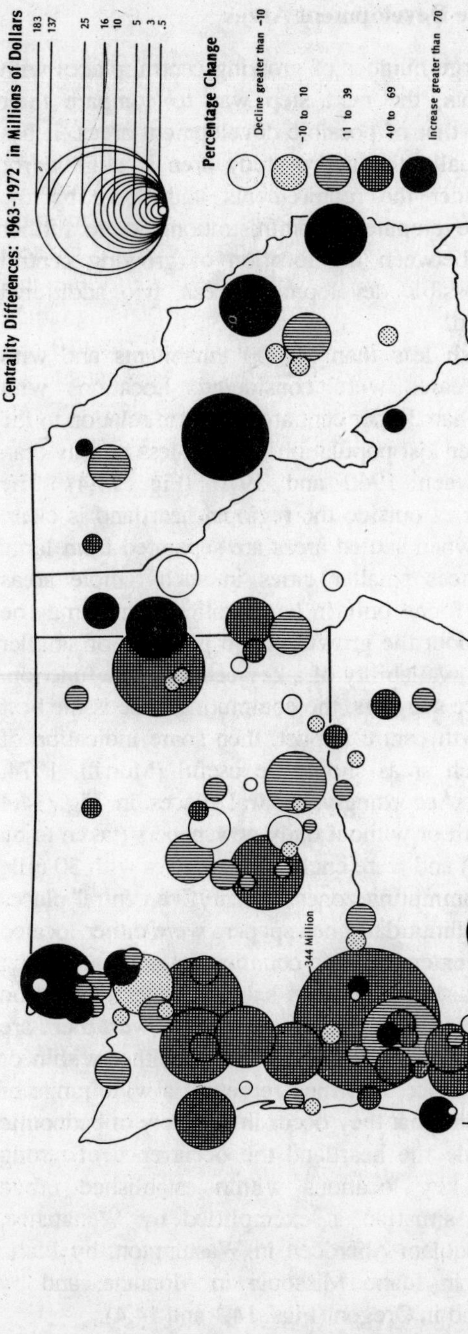

FIG. 14.3: Centrality differences by central place, 1967—72.

Central Places and Possible Development Areas

Given the existence of a large number of growing central places with less than 50,000 inhabitants, the next step was to compare their territorial arrangement with that of possible development areas. It has been pointed out that virtually the entire study area is eligible for development assistance under the requirements laid down by the United States Economic Development Administration. So, to further highlight the relationship between the location of growing central places of all sizes and possible development areas, two additional perspectives were established.

First, central places with less than 50,000 inhabitants and with exceptional centrality increases were considered. Locations with centrality increases greater than 36 per cent are shown in relation to the pattern of counties that either lost population or grew less rapidly than the national average between 1960 and 1970 (Fig. 14.4). The concentration of such counties outside the regional heartland is clear. Moreover, if it is true that when settled areas are separated from large cities by substantial distances smaller cities in such remote areas provide functions normally found only in larger cities, than it may be possible to tell still more about the growth centre potential of smaller central places by noting the availability of a key-central place function. And, if as available evidence suggests, the commuting zone is the best index of the extent of growth centre impact, then some indication of the territorial extent of such areas should be useful (Morrill, 1974, 1975; Berry, 1973, 1979). Accordingly, central places in Fig. 14.4 were divided into places with or without daily newspapers (taken to be a key central place function) and were enclosed by circles with 30 mile radii to indicate possible commuting zones. Twentyfive central places with daily papers and 10 without daily newspapers were either located in or had commuting zones extending into counties not only qualifying for federal development assistance but that showed either population loss or slow growth. It is thus apparent that in the North-west there are numerous economically strong central places located either within or close to areas needing assistance, that they represent a wide range of sizes and locational types, and that they occur in a variety of economic situations. Moreover, outside the heartland the occurrence of strong central places occupying key locations within established urban systems is frequent. This situation is exemplified by Wenatchee, Yakima, Tri-Cities and Hoquiam-Aberdeen in Washington, by Idaho Falls and Couer d'Alene in Idaho Missoula in Montana, and by Roseburg, Modford and Bend in Oregon (Figs. 14.1 and 14.4).

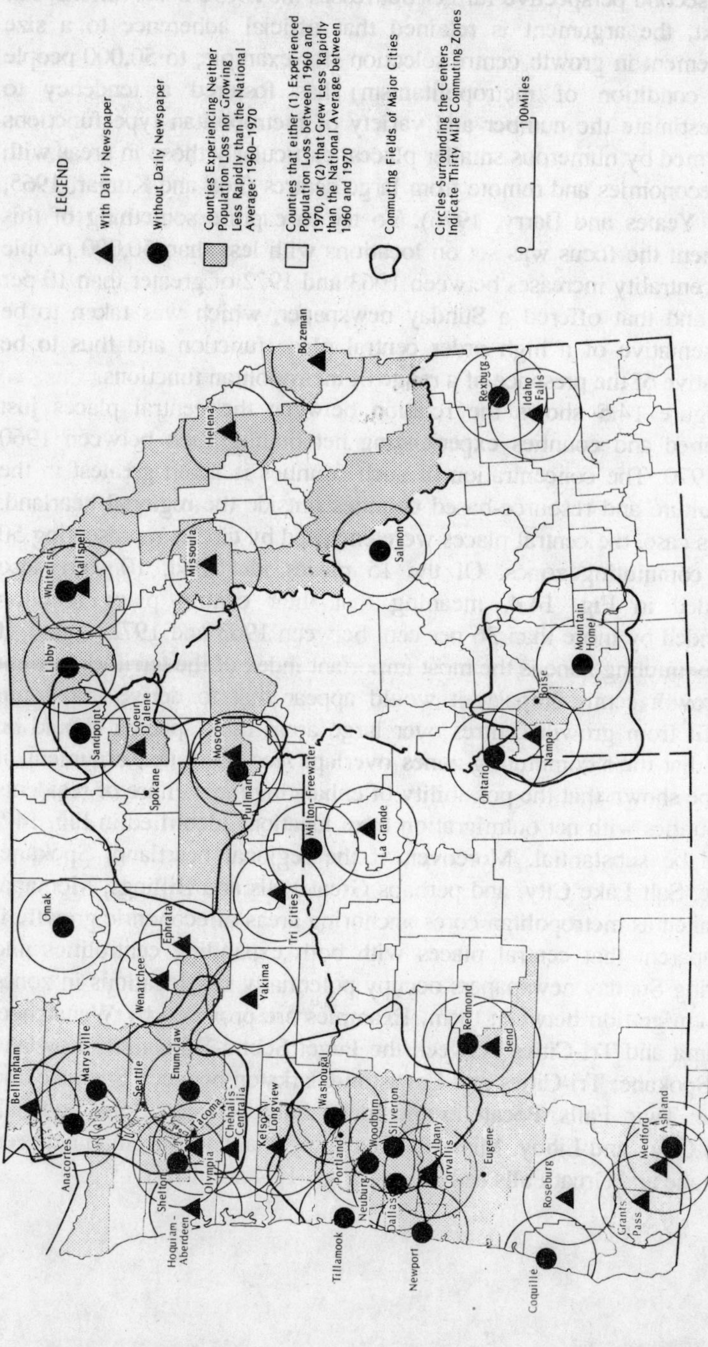

FIG. 14.4: Central places with populations less than 50,000 and centrality increases greater than 36 per cent, 1963–1972.

A second perspective further buttresses the above observations. For context, the argument is retained that official adherence to a size requirement in growth centre selection (for example, to 50,000 people as a condition of metropolitanism) has fostered a tendency to underestimate the number and variety of metropolitan type functions performed by numerous smaller places, particularly those in areas with rural economies and remote from large centres (Fox and Kumar, 1965, 1966; Yeates and Barry, 1976). To try to capture something of this argument the focus was set on locations with less than 50,000 people with centrality increases between 1963 and 1972 of greater than 10 per cent, and that offered a Sunday newspaper, which was taken to be representative of a high order central place function and thus to be indicative of the presence of a range of metropolitan functions.

Figure 14.5 shows the relation between the central places just described and counties experiencing net outmigration between 1960 and 1970. The concentration of such counties is again greatest in the agriculture and resource-based counties outside the regional hearland. In this case, the central places were enclosed by circles representing 50 mile commuting zones. Of the 15 places identified, 10 were also included in Fig. 14.4, meaning that their central place function expanded by more than 36 per cent between 1963 and 1972. Again, if the commuting zone is the most important index of the territorial extent of growth centre impact, it would appear that to derive maximum benefit from growth centres over large areas their spacing should be such that their commuting zones overlap. Accepting that argument, it will be shown that the possibility of enhancing the welfare of residents in counties with net outmigration from locations identified in Fig. 14.5 could be substantial. Moreover, if the regional heartland, Spokane, Boise, Salt Lake City, and perhaps Great Falls and Billings, Montana, are taken as metropolitan cores anchoring areas of economic growth, it is apparent that central places with both expanding centralities and offering Sunday newspapers occupy potentially key locations in zones of outmigration between them. Examples are provided by Wenatchee, Yakima and Tri-Cities between the Puget Sound-Willamette Lowland and Spokane; Tri-Cities and Lewiston-Clarkston between Spokane and Boise: Twin Falls, Pocatello and Idaho Falls between Boise and Salt Lake City : and Libby, Missoula, Butte, Helena, and Bozeman between Spokane and Great Falls or Billings (Figs. 14.4 and 14.5)

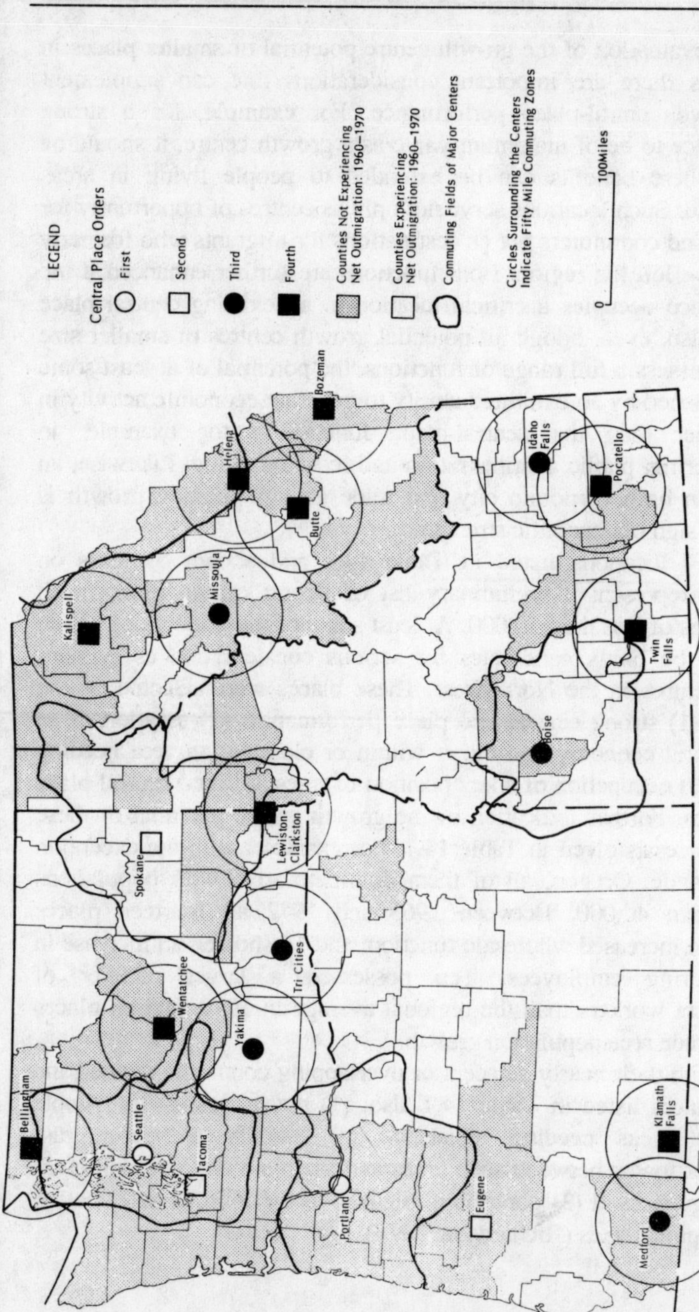

FIG. 14.5: Central places with populations less than 50,000 with Sunday newspapers and with centrality increases greater than 10 per cent, 1963–1972.

In an evaluation of the growth centre potential of smaller places in rural areas there are important considerations that can supplement comparative central-place performance. For example, for a strong central place to be of maximum value as a growth centre, it should be located where benefits can be extended to people living in areas needing aid. Such locations serve not only as centres of opportunity for residents and commuters but as destinations for migrants who formerly would have left the region. Both functions are further enhanced if the central place occupies a critical location in an existing central place system. Also, even though all potential growth centres of smaller size will not possess a full range of functions, the potential of at least some is strengthened by an exhibited ability to generate economic activity in areas other than the central-place function : for example, in manufacturing, public administration and in wholesaling. Likewise, an increase in both corporate city and trade area population growth is usually a sign of economic strength.

The 14 locations listed in Table 14.1 and shown by stars on Fig. 14.6 represent a preliminary list of strong central places with populations of less than 50,000. At least some of these locations would appear to be likely candidates for serious consideration as primary growth centres in the North-west. These places were selected on the basis of: (1) strong central and place performance, (2) location in an area of rural economy and either within or close by an area needing aid, and (3) occupation of a key position in an established central place sub-system. Further indication of the growth centre potential of these central places is given in Table 14.1. For example, with the exception of La Grande, Oregon, all of them dominate areas with populations greater than 40,000. Between 1963 and 1972 all fourteen places showed an increased wholesale function, and 11 showed an increase in manufacturing employees. Ten possessed a larger number of government workers than the regional average in 1970, and all places showed trade area population growth.

Through their nearly adjacent or overlapping commuting zones, the central places listed in Table 14.1 also: (1) reach most of the people living in areas needing assistance, (2) quality as out-migrant interceptor towns between such areas and the main metropolitan areas of the regions, and (3) contribute mightily to the long-term formation of development axes (Richardson, 1978).

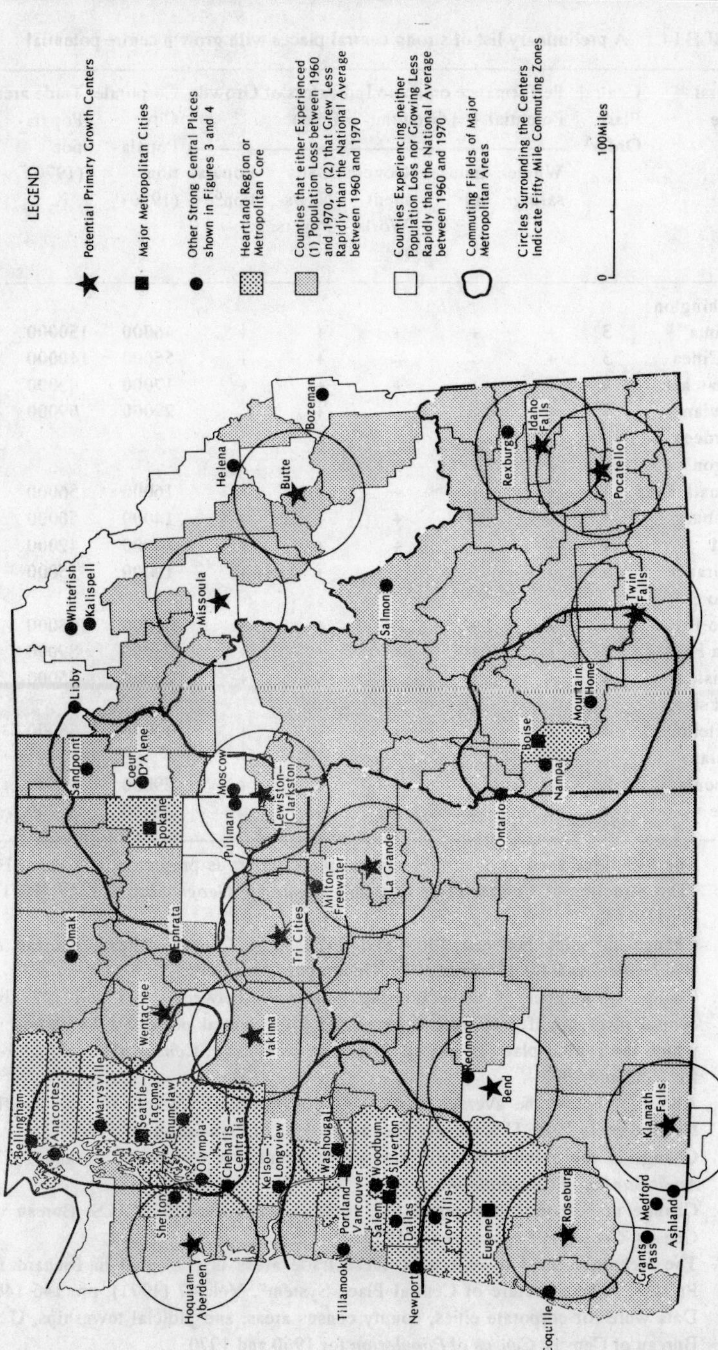

FIG. 14.6: A preliminary selection of strong central places with population less than 50,000 that show growth centre potential.

TABLE 14.1: A preliminary list of strong central places with growth centre potential

Central Place	Central Place Order[1]	Performance on some Indicators of Growth Potential(+ = Growth, - = Loss)					Corporate City Population (1970)	Trade area Population (1970)[7]
		Wholesaling[2]	Industry[3]	Government Workers[4]	Daily News-Papers[5]	Population[6]		
Washington								
Yakima	3	+	+	+	+	+	46000	150000
Tri-Cities	3	+	—	—	+	+	55000	140000
Wenatchee	4	+	+	+	+	+	17000	83000
Hoquiam-Aberdeen	4	+	—	—	+	—	29000	69000
Oregon								
Klamath Falls	4	+	+	+	+	—	16000	56000
Roseburg	4	+	+	+	+	+	14000	50000
Bend	4	+	+	+	+	+	14000	42000
La Grande	5	+	+	+	+	+	10000	15000
Idaho								
Idaho Falls	3	+	+	—	+	+	36000	103000
Twin Falls	4	+	+	+	+	+	21000	89000
Lewiston-Clarkston	4	+	+	—	+	+	26000	75000
Pocatello	4	+	+	+	+	+	40000	65000
Montana								
Missoula	3	+	+	+	+	+	29000	85000
Butte	4	+	—	+	+	—	23000	85000

1. The technique used to determine central place orders is presented in Richard E., "The Structure of Central Place System", *Economic Geography,* Vol. 47 (1971), pp. 140-142.
2. Change in sales between 1963 and 1972. U.S. Census Bureau, *Census of Wholesale Trade* for 1963 and 1972, Geographic Area Series.
3. Change in number of manufacturing employees between 1963 and 1972 by central place, or, if data were not reported for a central place, by the county in which the central place is located. U.S. Census Bureau, *Census of Manufactures* for 1963 and 1972.
4. Above or below the average number of governmental workers for the Pacific North-west in 1970, U.S. Census Bureau, *Census of Population* for 1970.
5. Change in the volume of daily newspaper circulation between 1963 and 1972. *Audit Bureau of Circulations.*
6. Change in corporate city population between 1960 and 1970. U.S. Bureau of Census, *Census of Population* for 1970.
7. The technique used to determine these trade areas is presented in Richard E. Preston, "The Structure of Central Place System", Vol. 47 (1971), pp. 146-148. Data were for corporate cities, county census areas, and judicial townships, U.S. Bureau of Census, *Census of Population* for 1960 and 1970.

The principal axis is already well developed between Vancouver, British Columbia and Medford, Oregon. A second axis is forming between Seattle-Tacoma and Spokane, and an extension is emerging between Spokane and Billings, Montana, by way of Missoula, Butte and Bozeman. Another axis is developing between Boise and Idaho Falls, and one day may extend to the urban complex centred on Salt Lake City. Other corridors that may develop over the long-term run north-south along the eastern base of the Cascade Mountains (through Wenatchee, Ellensburg, Yakima, Tri-Cities and Bend), and between Tri-Cities and Boise by way of Pendleton, La Grande and Baker.[11]

Also included in Fig. 14.6 are the remaining places with strong central-place functions shown in Figs. 14.4 and 14.5. In several cases they occupy critical locations near the outer edge of commuting zones of either Seattle-Tacoma, Portland-Vancouver, Boise, or Spokane (for example, Coeur d' Alene, Sandpoint, Libby and Pullman-Moscow in Spokane's field) or of potential growth centres (for example Omak and Ephrata in Wenatchee's field). Such small but central places not only contribute to both the overall strength of the daily urban systems and development corridors in which they are located, they have potential as secondary growth centres from which opportunities present in the economically robust daily urban systems in which they are located could be extended still farther (at least to the outer reaches of the secondary centre's commutation zone) into needy areas beyond the boundaries of a primary growth centre's commuting range.

Conclusions and Implications

It has been established in this study that despite the qualification of virtually all Pacific North-west counties for federal development assistance, the presence of vast areas showing either population decline or slow growth between 1960 and 1970, and the presence of extensive areas showing net outmigration between 1960 and 1970, that the region possesses numerous strong and growing central places. On the basis of the argument presented, therefore, it appears that some or all of the places shown in Table 14.1 could provide strong rallying points for development of less favoured sections of the North-west. Additionally, since numerous central places representing a variety of sizes, locations and local economies show strong growth without explicit government assistance, it follows that through judicious selection and assistance procedures, growing (or in some cases stable) central places representing a range of sizes and locations even wider than those suggested here could serve as either primary or secondary growth centres in appropriate rural development schemes in the North-west. At a broader level support is offered here for Morrill's suggestions that : (1) flexibility is possible in planning growth centre locations, (2) that it is not necessary to insist upon either a particular size or spacing (these can vary according to regional conditions), and (3) that a search "should be directed not at finding cities of a particular minimum size, but at finding firms that can utilise local labour and compete successfully in regional and national markets" (Morrill, 1973a).

The argument offered here is embryonic and the actual list of potential growth centres tentative, but the evidence appears to be sufficiently encouraging to merit both further investigation and consideration in formal growth centre selection procedures aimed at smaller places. In closing, two groups of observations are offered. First, present approaches to the determination of growth centre potential largely ignore smaller places because convincing methods have not been produced for evaluating small centre prospects. Towards a solution to his problem it was suggested and demonstrated that classical central place ideas offer a point of departure for differentiating the growth centre potential of such places. Specifically, it is suggested that knowledge of the magnitude of each location's central place function and of its pattern of recent change can be critical in such evaluations. In this regard, it was also suggested that a stable or expanding central place function might stand as a solid base for employment increments consistent with the scale of a central place. This approach could help identify growth centres representing a wide range of urban sizes and locations and perhaps help to ensure viable settlement systems in parts of North America lacking large cities.

Second, several other contributions to a solution for the problem confronted here are offered by a central place perspective. Because central place ideas focus squarely on tertiary and institutional activities, a base is identified by which places of virtually any size can be compared. Concentrating on the central place function identifies the relative strength as regional centres of the main places in every settlement sub-system in a region and thus draws attention to the place in each sub-system that has developed as the focus of interaction. The systems view offered by central place ideas reveals places likely to benefit from commuting opportunities, migration possibilities, or by employment decentralisation. When such systems are identified by communication linkages like daily commutation and newspaper circulation, territories are identified where there is a likelihood that knowledge of opportunity in a central place could spread and thus facilitate what Richardson calls the socio-economic function of growth centres (Richardson, 1969). Central places, particularly those remote from large cities, are also likely to offer a range of goods and services unusual for places of their population size, and thus provide a stronger base for potential development than places with equal populations but with a weaker central place roles. Finally, central place ideas address those sectors of the economy where the rate of job formation is presently greatest. In summary, classical central place ideas may offer a practical contribution to the solution of what is now an open problem facing regional development programmes employing growth centre strategies.

NOTES

1. A strong literature does not exist on the problem of growth centre selection. Malcom J. Moseley. *Growth Centres in Spatial Planning* (London : Pergamon Press, 1974), pp. 10. 164-165. Niles M. Hansen. *Rural Poverty and the Urban Crisis* (Bloomington : University of Indiana Press, 1971). p. 301. The nature of growth centre-surrounding area relations are also poorly documented, but there is a solid theoretical perspective emerging, Richard L. Morrill, "Growth Centre-Hinterland Relations", in F. Helleiner and W. Stohr (eds.), *Spatial Aspects of the Development Process.* Vol. II of the Proceeding of the Commission on Regional Aspects of Development of the International Geographical Union (Toronto : Allister Typesetting and Graphics, 1974), pp. 215-240; and R. L. Morrill "Impacts of Urban Growth Centres on their Hinterlands", *Urban-development in the USA and Hungary* (Budapest : Akademiai, Kiads, 1975), pp. 52-72.
2. Although all reviews of the growth centre concept include sections on Central Place Theory, two of the most complete discussions are by John B. Parr, "Growth Poles, Regional Development and Central Place Theory, *"Papers of the Regional Science Association,* 31 (1973), pp. 173-212, and Tormod Hermansen, "Development Poles and Related Theories : A Synoptic Review", in Niles M. Hansen (ed.), *Growth Centres in Regional Economic Development* (New York : The Free Press 1972), pp. 160-203. In this study, the terms "Central Place Theory" and "Classical Central Place Theory" refer to the contribution of Walter Christaller, *Central Places in Southern Germany,* trans, by Carlisle W. Bakin (Englewood Cliffs, N.J. : Prentice Hall, 1966), and not to the recent work known as "Modern Central Place Theory". Christaller offered a partial theory of the space economy dealing with the distribution of trades and institutions at the inter-city scale. The Modern Central Place Theory attempts to generalise and thus broaden the application of Christaller's work (to the intra-city scale, for example) and is presented in Brian J. I. Berry, *Geography of Market Centres and Retail*

Distribution (Englewood Cliffs, N.J. : Prentice Hall, 1967), pp. 1-88 and Keith S.O. Beavon, *Central Place Theory, A Reinterpretation* (London Longmans, 1977).

3. For traditional approaches to this problem see, Moseley, op. cit. Hansen, *Rural Poverty and the Urban Crisis,* op. cit. Niles M. Hansen, *Intermediate-Size Cities as Growth Centres* (New York : Prager Publishers, 1971), Harry W. Richardson, "Growth Centres, Rural Development and National Policy : a Defense". *International Regional Science Review,* 3 (1978), pp. 133-152.
4. No attempt is made to specify a minimum viable size for growth centres, An introduction to this and associated issues is given in Moseley, op. cit. pp. 74-75, 113, 145-148, 160-161.
5. For an example of the first point, see, Brian, J. L. Berry, *Growth Centres in the American Urban System,* Vol. 1 (Cambridge, Mass. : Billinger Publishing Company, 1973). On the second point see, Harry W. Richardson, *The Economics of Urban Size* (Lexington, Mass. : Lexington Book, 1973), pp. 181-184, and Moseley, op. cit., pp. 17-18. The general case for smaller places as growth centres is best presented by Richard L. Morrill, "On the Size and Spacing of Growth Centres", *Growth and Chance,* Vol. 4 (1973), pp. 21-24, and in Richard L. Morrill, "Fundamental Issues Concerning Future Settlement in America", in *Geographic Perspectives and Urban Problems,* edited by F.E. Horton, (Washington, D.C. : National Academy of Sciences, 1973), pp. 81-93.
6. As population continues to concentrate in "urban fields", and as economic activities continue to disperse within such areas, it would appear that the role of the classical Central Place Theory in the description and explanation of the structure of such regions will increase.
7. For both the conceptual and operational definitions of centrality, see Christaller, op. cit., pp. 17-18, 137-153.
8. For the development and at least partial verification of this model, and for its relationship to Christaller's interpretation of centrality, see Richard E. Preston, "Two Centrality Models", *Yearbook of the Association of Pacific Coast Geographers,* Vol. 32 (1970), pp. 59-78, and Richard E. Preston, "Toward Verification of A Classical Centrality Model", *Tijdschrift Voor Economishe en Sociale Geographie,* Vol. 62, (1971), pp. 301-307, and Compare with Christaller, op. cit., pp. 17-18, 137-153.
9. The idea of considering several places within a single labour market as one location to capture size benefits will undoubtedly prove useful in the designation of locations as growth centres, particularly in areas with few large cities, Moseley, op.cit., p. 113.
10. Although the combined corporate city populations of the Tri-Cities of Richland, Kennewick, and Pasco, Washington, totalled 55,000 in 1970.
11. Both the established axes and the incipient ones are clear from the map of highway traffic on primary roads (1960-70), in Stephen S. Birdsall and John W. Forin, *Regional Landscapes of the United States* (New York : Wiley, 1978). p. 27.

REFERENCES

Beavon, K. S. O., 1977 : *Central Place Theory: A Reinterpretation,* London.

Berry, B. J. L., 1967 : *Geography of Market Centres and Retail Distribution,* Englewood Cliff, pp. 1-88.

———, 1970 : "The Geography of the United States in the Year 2000", *Transactions of the Institute of British Geographers,* No. 57.

———, 1973 : *Growth Centres in American System,* Vol. I, Cambridge.

———, 1976 : "The Counter-urbanisation Process: Urban America Since 1970", in Berry B. J. L. (Ed), *Urbanisation and Counter-urbanisation,* pp. 17-30.

———, 1979 : "Labour Market Participation and Regional Potential", *Growth and Changes,* Vol. IV, pp. 3-10.

Birdsall, S. S. and Forin, J. W., 1978 : *Regional Landscape of United States,* New York, p. 27.

Christaller, W., 1966 : *Central Places in Southern Germany,* trans. by Carlisk, W. Bakin.

"Economic Development Districts : January 1, 1978" (Washington D.C. : U.S. Department of Commerce, Economic Development Administration, 1978). "EDA Qualified Areas, July 1978" (Washington D.C. : U.S. Department of Commerce, Economic Development Administration, 1978).

Fox, K. A. and Kumar, T. K., 1965 : "The Functional Economic Area : Delineation and Implications for Economic Analysis and Policy", *Regional Science Association Papers,* Vol. 15, pp. 57-86.

———, 1966 : Delineating Functional Economic Areas", in *Research and Education for Regional and Area Development,* pp. 13-55.

Friedmann, J. and Weaver, C., 1978 : *Mount on Territory and Function,* London.

Hansen, N. M., 1971a : *Rural Poverty and the Urban Crisis,* Bloomington, p. 301.

———, 1971b : *Intermediate Size Cities and Growth Centres,* New York.

———, 1975 : "An Evaluation of Growth Centre Theory and Practice", *Environment and Planning,* Vol. VII, pp. 821-832.

Hermansen, T., 1968 : "Service Trades and Growth Centres", in *Regional Policy in EFTA : An Examination of the Growth Centre Idea,* Edinburgh, pp. 143-209.

———, 1972 : "Development Poles and Related Theories : A Synoptic Review", in Hansen, N. M. (Ed), *Growth Centres in Regional Economic Development,* New York, pp. 160-203.

Highsmith, R. J. (Ed.), 1962, 1973: *Atlas of the Pacific North-West.*

Hoover, E. M., 1975 : *An Introduction to Regional Economics,* New York, pp. 277-283.

Kuklinski, A., 1972 : *Growth Poles and Growth Centres in Regional Development.*

Morrill, R. L., 1973a : "On the Size and Spacing of Growth Centres", *Growth and Chance,* Vol. IV, pp. 21-24.

———, 1973b : "Fundamental Issues Concerning Future Settlement in America", in Horton F. E. (Ed.), *Geographic Perspectives and Urban Problems,* Washington, pp. 81-93.

———, 1974: "Growth Centre Hinterland Relations", in Hellenier, F. and Stohr, W. (Eds), *Spatial Aspects of the Development Process,* Vol. II, Toronto, pp. 215-40.

———, 1975 : "Impact of Urban Growth Centres : On their Hinterlands" *Urban Development in U.S.A. and Hungary,* Budapest, pp. 52-72.

Moseley, M. J., 1974 : *Growth Centres in Spatial Planning,* London, pp. 10, 164-165.

Office of the Chief Council, "Public Works and Economic Development Act of 1965. As Amended" (Washington D.C., : U.S. Department of Commerce, Economic Development Administration, 1976), Pacific North-west Regional Commission, *Regional Plan* (Vancouver, Washington : Pacific North-west Regional Commission,

U.S. Department of Commerce, 1977).

Pacific North-west Regional Commission, op. cit., Richard J. Highsmith (ed.), *Atlas of the Pacific North-west* (Corvallis : Oregon State University Press, 3rd and 5th editions, 1962, 1973). Richard E. Preston: "The Structure of Central Place Systems", *Economic Geography,* Vol. 47, (1971), pp. 136-155.

Pars, J. B., 1973 : "Growth Poles, Regional Development and Central Place Theory", *Papers of the Regional Science Association,* pp. 173-212.

Preston, R. E., 1971a : "The Structure of the Central Place Systems", *Economic Geography,* Vol, 47, pp. 136-155.

———, 1970 : "Two Centrality Models", *Yearbook of the Association of Pacific Coast Geographer,* Vol. 32, pp. 59-78.

———, 1971b : "Towards Verification of a Classical Centrality Model", *Tijds-chrift Voor Economische en Sociale Geographie,* Vol. 62, pp. 301-307.

Richardson, H. W., 1969 : *Regional Economics,* New York, p. 425.

———, 1973a : *The Economies of Urban Size,* Lexington, pp. 181-184.

———, 1973b : *Regional Growth Theory,* London, p. 50.

———, 1978 : "Growth Centres, Rural Development and National Policy. A Defence", *International Regional Science Review,* pp. 133-152.

Simmons, J. W., 1976 : "Short-term Income Growth in the Canadian Urban Systems", *Canadian Geographer,* Vol. XX, pp. 419-431.

Yeates, M. and Barry, G., 1976 : *The North American City,* New York, p. 40.

15

MODERN AGRICULTURE AND RURAL SETTLEMENT

GYORGY ENYEDI

PRIOR TO World War II Hungary was an agrarian-industrial State with a poorly developed industrial base. According to the census of 1941, 50 per cent of the active population was employed in agriculture and 60 per cent of the total population lived in the rural settlements. As a result of the speedy urbanisation and industrialisation since 1945, today industry accounts for 70 per cent of Hungary's GNP and 51 per cent of the population live in the cities. Since our economic geography and settlement science specialists have also been primarily concerned with this rapid urban-industrial development, there was scant attention paid to research concerning the rural area. The development of the settlement network and regional planning was also urban-centred. Consequently, the inadequately planned and often spontaneous transformation of the rural settlement network is not known sufficiently.

The importance of the rural sphere, however, can scarcely be doubted — 49 per cent of our population and 50 per cent of all industrial workers still live within it and the total population of the rural area has decreased only by 3-4 per cent during the last four decades. The importance of agriculture — the primary function of the rural sphere — has increased as a result of the development of extensive foreign trade activities and of the strategical and economic role of food production.

The definition of "rural" and "urban" spheres varies from country to country and in general they reflect the historical development of the settlement network of a given State. In Hungary the village and the city are primarily administrative units and the statistical concepts of "urbanity" merely follow the administrative lines. The size of the population is not a primary criterion: in the Hungarian Great Plain there are rural settlements (villages) with 15,000–20,000 people residing within them. Generally, a settlement becomes officially recognised as a city considerably later than the functional existence of urban networks. Administratively there are only 95 cities in Hungary, even though, according to the findings of settlement geographers, 140–150 settlements fulfil urban functions.[1]

In the Geographical Institute of the Hungarian Academy of Science, research has been concentrated on the transformation of the villages of Hungary since 1976. In this context the rural sphere has been defined as that area which, alongwith its natural resources, settlements and population, falls outside the urban agglomerations. Thus even small towns existing outside such agglomerations are regarded as part of the rural sphere.

The general characteristics of the transformation of the rural sphere that had taken place during the last 30 years are as follows :

1. The size of the rural population has decreased only minimally, within the rural sphere one could witness major population movements. Significant decreases occurred in the non-industrialised tiny villages of south-western Hungary and the Northern mountainous region where there are no major urban centres and where the natural agrarian environment is unfavourable.

   At the same time, the population has increased in the commuting zones of the cities, especially around the north-east, south-west industrial axis, and in those villages with more than 5,000 inhabitants that fulfil some urban functions.

   The intra-State migrations decreased and such migrations as do take place are no longer unidirectional toward the cities. Between 1970 and 1975 the permanent migrants moved from one village to another and there had been labour movement among the various agrarian production units.
2. The socio-economic structures of village and city have become similar to each other, while, naturally, only 10-12 per cent of those employed in agriculture live in the cities and the percentage of those employed in the service sector is much higher there. It is significant that the ratio of industrial workers is the same in the city and the village. Unlike 30 years ago when

most people living in the village were agricultural employees, today only 38 per cent of the rural population is employed in agriculture. Moreover, the gap between industrial and agricultural incomes had been eliminated during the last 10 years. A fundamental feature of their social equalisation has been the basic transformation of the agricultural employment.

3. The previously predominant agrarian character of the rural sphere has become multi-functional. Especially between 1960 and 1975 significant industry moved into the villages. At present around one-fifth of the factory workers — around 400,000 individuals — work in factories located in the villages and this figure does not include those working in the industrial production branches of the agricultural cooperatives. Even the residential function of the rural settlements expanded significantly — more than one million workers commute daily from rural residences to urban workplaces. In addition the tourists and environmental protection functions of the rural sphere have also been on the increase.
4. The modernisation of rural functions has been faster than the modernisation of rural settlement infrastructure. Consequently, tensions have been evident between the speedy socio-economic development and the level of rural existence. On the one hand, progress was evident : in 1963 all the villages had been supplied with electricity, in 1974 they were all connected with the national road and bus transportation network, by 1978, 60 per cent of the rural homes cooked with gas and 40 per cent had running water. On the other hand, the level of rural existence was significantly below that of the city infrastructurally and in regard to services thus in Hungary at the end of the 1970s a chief source of social inequality remains the difference between "town" and "country".

The transformation of the Hungarian village took place as a result of such exogenous factors as the general process of urbanisation and of indigenous factors, namely, the general modernisation of agriculture in Hungary. In spite of the decrease in the agricultural population, agriculture remains the most important rural function : 75 per cent of the land remains under agricultural and a further 15 per cent under forestry use. The annual growth index of food production since 1970 is hardly below that of industrial production and, according to the 1978 report of the World Bank, between 1967 and 1977 Hungary's agricultural growth had been exceeded only by the Netherlands.

Development Phases of Modern Agriculture

One can discern several somewhat delineable phases in the development of the modern agriculture. The *first* is the period of mechanisation and economic concentration that creates the *technical precondition* for large-scale agricultural production. The *second* phase is the period of the so-called *biological revolution* that creates the preconditions for the large-scale, serially produced, similar quality produce through the introduction of higher plant and animal yields. The *third* phase is characterised by the fundamental modernisation of economic organisation, during which agricultural production is suffused with industrial production and merchandising techniques. The chief characteristics of the agricultural transformation of the 1970s and 1980s is this organisational change.

The most significant aspect of this organisational change is the integration of previously sharp economic activities. While in traditional agriculture the production of most of the tools necessary for that production, that of the produce and the sale of the end-product as well was done by the peasantry itself, today four groups of produces must harmonise their activities to ensure the production and merchandising of agrarian goods. These are :

(a) the service branches that ensure supplies and means of production;
(b) the so-called technical-intellectual branches that ensure the development of such services as research, consulting and specialist training;
(c) the agricultural production itself; and
(d) the preparation and sale of the final agricultural produce.

The integration of the developed socialist and capitalist large-scale agriculture is fundamentally different. In the capitalist system through the integration process the functions of the farm decrease, their operation more and more is directed from the outside and the role of the integrators is played by the large agricultural-commercial firms. The family farm is increasingly dependent upon the industrial and commercial organisations existing above the level of the farm and becomes a cog in the wheel of production. The farmer remains independent but the control over and the decisions about his own production activities slip out of his hands. Thus, for example, in the past the complete organisation of poultry and egg production was in the hands of the farmer who had selected his stad, grew his chickens and sold his eggs on the market all by himself. But, as a result of the integration process, the breeding, the fodder as also the merchandising have been taken out of his hands; the mammoth agricultural firms today decide upon the proper type of chicken and egg hybrids, preparation of fodder and forage and the sale of the eggs. The production of food has been diffused geographically as well as organisationally; storage, preparation, feed production have been separated from the production locus and have been concentrated largely in the cities.

In the capitalist agricultural-industrial integration the comparative importance of agriculture has been sliding — in 1972, for example, in France the agricultural share of food products from the final price was only 28 per cent — and the concentration of the large firms and multi-nationals has been on the increase.

In contrast the integration of large-scale socialist agriculture has been characterised by the following general processes :

(a) the large-scale farms — whether state or cooperative — can conduct their activities on an equal basis with the commercial or food-preparation firms. As a result of the small size of the farm, there are very few farms — such as the King's Ranch — in the capitalist system which can afford to do that. Lest one forget an average-sized Hungarian cooperative is 15 times larger than an average-sized American farm. The large farms in the socialist system create joint companies with the State food-industrial or large-scale commercial firms; the levels of integration become blurred, but the agricultural unit does not become subordinate to higher levels;

(b) the large-scale farms produce such enormous amounts that they engage at least partially in their preparation and sale;

(c) the development and breeding of new branches, hybrids, etc., the employment of new production processes and the innovation in general are generally tied to State farms and to a few special cooperatives, and, therefore, it remains within the agricultural production branch itself. Hence, through the process of horizontal integration, for example, in the employment of various production systems, various economic units of the same level come into close contact with one another;

(d) the processes described above have decentralised a large number of industrial, transportation and commercial activities in the rural sphere in the large agricultural firms. As a result, a large part of the population listed in statistical indices is employed in agriculture, in reality, are engaged in non-agricultural work; and

(e) finally, while the integration in the developed capitalist States results in a functional impoverishment and a consequent further loss of prestige of the village, the large-scale economic activities taking place in the socialist State generally stabilise rural life, broaden rural activities and sometimes level dynamism to the locus as well.

Effects of Modern Agrarian Development

The technical-organisational change briefly sketched above, naturally, has had significant influence upon the rural settlements. Here attention will be concentrated on the influence upon the settlement network itself, without regard to the internal settlement structure of the village. The most important effects of modern agrarian development had been the following :

Internal Transformation of Agricultural Population Itself

The concept of agricultural population completely changed during the course of collectivisation : the old traditional peasantry nearly completely disappeared from the Hungarian villages. In the large-scale agricultural units traditional peasant labour — based on familial division of work and universal type of agrarian activity — was no longer needed. The breakdown of labour processes into smaller component parts made technical adaptation easier, although as an unwanted result today's agricultural skilled labourer is less able to see the fruit of his labour.

In 1973 the vast majority — 70.5 per cent — of agricultural employees worked in cooperatives, 15.2 per cent in State farms, 5.1 per cent in fisheries, 4.2 per cent in forestry and 5 per cent worked on their own private lands. The cooperatives also undertook significant industrial activities: one-third of their total income came from non-agricultural activities. Thirty per cent of all cooperative members and employees were engaged in these industrial, construction, commercial, trade or warehouse activities. Moreover, of the roughly 18 per cent, who are officially listed as being engaged in "agricultural" work, as many as 12 per cent were engaged in white-collar, administrative or management activities. Thus, these 18 per cent cannot be compared on a one-to-one basis with agricultural labourers engaged in private economic activity.

A special characteristic of Hungarian agriculture is the enormous importance of the auxiliary farming. In this type of activity the traditional familial division of labour continues to predominate, hence the sociologists refer to the dominating life-style as that of the "post-peasants". Although originally intended to serve the food self-sufficiency needs of the cooperative peasantry, by now its importance has vastly exceeded the presumed goals.

The majority of the household plots are owned by non-agricultural employees : 40 per cent of all Hungarian families are engaged in these activities. Thirty per cent of all income derived from the household plots is accrued to the working class, 23 per cent to the peasantry and the remainder is divided among the rest of society. The auxiliary farms (or household plots) are more and more used to produce for the market and by now more than 50 per cent are sold there. In 1975, 36 per cent of gross agricultural production value came from the household plots that comprise only 14 per cent of all agricultural land.

The household plots are engaged in the most intensive type of production activities : they offer 50 to 55 per cent of all hog, poultry, viticultural, fruit and vegetable production. The households have exceedingly close ties with the large-scale farms, buy their fodder and lease machines from them, and thus the household farm can be regarded as an integral component of socialist agriculture. The household plots save a large amount of capital invested for the State through, for example, the use of available stables and also utilise excessive labour. According to some calculation 2.7 billion man-hours were spent on the household in 1972-73, only 30 per cent less than on the large-scale units. Sixty one per cent of these 2.7 billion man-hours, however, were worked by dependents or those in retirement.

Among the agricultural employees, those engaged in skilled labour and industrial or service activities continue to demand greater services and better technological, infrastructural environment in the village. Since the consumption habits of rural and urban residents are beginning to be similar to each other, rural residents begin to demand services similar to those found in cities. The great importance of the household plots acts as a conservative force: it retains at least partially the productive function of the rural residence and through this maintains the ground plan of the rural settlement as well.

New Ties between Large-Scale Farms and Rural Settlements

The growth of farm size had important effects upon the rural settlements. During the first half of the decade of the 1970, the State and cooperative farms were merged with neighbouring units in order to create more size-efficient farms. In 1960 there were 333 State and 4,507 cooperative farms : by 1975 their numbers were reduced to 151 and 1,599, respectively. Consequently in 1975 the average area of a State farm was 6,602 hectares, that of a cooperative farm 3,161 hectares. Thus, on more than one-third of the territory of the State — in areas characterised by small clustered rural settlements — a farm area includes several villages as well. While earlier a village provided a frame for the agricultural activity, now the farm holds together several villages in some of which there certainly are people who are employed in other sectors of the economy. The settlement development and economic development plans are not always sufficiently harmonised. Thus, for example, the unified cooperative does not always move to the administrative centre to share their centrifugal role of the local council. The development of the villages almost always is based on local efforts and sources and among the income sources the taxes laid on the collective and State farms are always the most significant. Indeed, it happens quite frequently that the cooperatives or State farms take the leading role — and build houses for their employees outside the existing villages.

During the course of the modernisation of agriculture the infrastructural sensitivity of agriculture itself had increased greatly. In our opinion, in the most productive regions of the Great Plain, the reason for the significant underutilisation of the available soil capacity is the underdeveloped infrastructure. Thus, for example, the demand of agriculture for electric energy, gas, telephone, telex, etc., has grown enormously, just as much as their needs for transportation networks (roads, etc.) within the villages. The raw products released by agriculture increase by 3.5 per cent annually and, according to forecasts, is expected to double by the turn of the century. Such growth cannot be expected of any other mineral product. The transportation demand of agricultural goods takes place unevenly in time and place, but continuously and everywhere. The improvement in the rural road network is thus demanded by the necessity of collecting and transporting the product as well as by the increased demands for human labour transportation within the large modern agricultural units. The modernisation of agriculture, therefore, can conceivably quicken the development of the rural infrastructure and can significantly contribute to the betterment of rural existence.

Development of Regional Ties within Rural Sphere

The modernisation of agriculture significantly increased the hitherto exceedingly thin network of contact among the various villages. It is important to note that the ties between city and village are no longer restricted to a trade or industry for agricultural goods or the utilisation of urban services. The agro-industrial integration tied the urban and rural settlements into a single economic chain: the food produced in the villages, the processing industries partly in the cities, the industries serving agriculture with machinery are all tied together in organised forms. Another aspect of urban-rural ties is derived from the fact the 15– 20 per cent of those agricultural specialists now reside in the cities and commute to the nearby villages.

Moreover, the organised and continuous ties among the villages have also taken place within the confines of modern agriculture. Naturally, the closest ties exist among those villages that are located under the umbrella of a common farm. Many types of economic cooperation have developed among the various neighbouring villages, for example, jointly owned slaughter-houses, soils, warehouses or processing plants have been created. The most complex ties among the villages, however, have been developed through the adoption of various production systems.

The production systems are designed to produce on a modern, industrial-like basis a single commodity or crop and serve as the most important basis of horizontal integration. The centre of the system is most often a State farm or an exceedingly good cooperative farm. The innovational centre, therefore, remains within agriculture. Between the centre of the system and its member units, the ties are continuous and well organised, since special assistance, consulting and control are also included as part of the centre's activity. These production systems are not developed regionally, and, therefore, partner members are tied to a system from vastly different parts of the State.

In 1977, 67 production systems were operated : of these 21 were crop production, 27 fruit and viticultural and 19 animal-husbandry systems. 86.6 per cent of the State and 78.3 per cent of the cooperative farms were members of one or more systems. One-third of the arable plough-land and one-fifth of the fruit and veneyards were worked by these modern systems and even in animal husbandry the modern system's management approach was extended to 50 per cent of all animals produced on a large scale.

Summary

The rural settlement had been harmonious with the surrounding agriculture; the size of the village, the topographical location of its population, its internal structure, its houses, the character of its services, the road network were all adapted to the needs of agriculture itself. This traditional harmony — precisely because of the slow development of agriculture and the stagnation of rural society — survived even in the first-half of the 20th century. The socio-economic development of the last 30 years had broken that traditional harmony once and for all. As a result of the rapid social transformation, the rural settlements must serve more than the needs of agriculture only. But the demands of agriculture have also been altered significantly as a result of the technological development and the agro-industrial integration during the last decade. The technical infrastructural development of the rural sphere was considerably behind that of agriculture: in fact, it was not even able to satisfy the needs of the latter. As a result tension has developed between the village and agriculture, the resolution of which is necessary both because of the needs for increased agricultural output and because of the need to improve the standard of living of the rural population. The rural settlement network attempts to adapt quite spontaneously to the changed social environment for which it provides a framework. It is high time to attempt to define the rural model and the rural-urban relationship of the developed socialist system, in order to allow our settlement planners to plan for the entire settlement network and not to restrict its development merely to our cities and urban spheres.

NOTE

1. If we were to employ the same definition as that which is used in the GDR, 80-82 per cent of Hungary's population would be identified as "urban". Hence we must be careful to utilise cross-national comparisons, lest we come to an "statistically" proper though absurd conclusion that more people live in urban areas in Bulgaria than in Czechoslovakia.

REFERENCES

Androka, R., 1974 : A Kozsegi nepesseg tarsadalmi jellemzoi (Social characteristics of the rural population), *Tarsadalmi Szemle*, Vol. 29, No. 8-9, p. 69-75.

Barta, Gy., 1978 : "Hungary's Rural Industry", in *Urban Development in the USA and Hungary*, Studies in Geography in Hungary No. 14, Akademiai Kiado, Budapest, pp. 293-301.

Beluszky, P., 1965 : Falusi telepuleseink osztalyozasa (A Classification of our Rural Settlements), *Foldrajzi Ertesito*, Vol. 14, No. 2, pp. 149-163.

Enyedi, Gy. : A magyar falu atalakulasa (The Transformation of the Hungarian Village), *Foldrajzi Kozlemenyek*, vol. 23, No. 2, pp. 109-124.

Enyedi, Gy., 1977 : A korszeru falufejlestes elvei as legujabb tendencial. (Principles and New Tendencies of the Modern Rural Development), *Teruliltrendezes*, Vol. 24, No. 4, pp. 11-15.

Enyedi, Gy., 1978 : A mezogazdasag fejlodesenek hatasa a falusi telepulesekre. (The Influence of the Agricultural Development upon the Rural Settlements), *Varosepites*, Vol. 18, No. 1, p. 2-4.

Lettrich, E., 1975 : Les traits characteristiques de I, urbanisation en Hongrie, In-*Urbanisation in Europe*, Akademiai Kiado, Budapest, pp. 133-137.

Sipos, A., 1977 : Az agrar-ipri integracio (The Agro-Industrial Integration), *Kozgazdasagi Szemle*, Vol. 24, No. 6, pp. 613-627.

INDEX